Edward Payson Evans

Evolutional ethics and animal psychology

Edward Payson Evans

Evolutional ethics and animal psychology

ISBN/EAN: 9783337229665

Printed in Europe, USA, Canada, Australia, Japan

Cover: Foto ©berggeist007 / pixelio.de

More available books at **www.hansebooks.com**

EVOLUTIONAL ETHICS
AND
ANIMAL PSYCHOLOGY

BY

E. P. EVANS

AUTHOR OF ANIMAL SYMBOLISM IN ECCLESIASTICAL
ARCHITECTURE, THE CRIMINAL PROSECUTION AND
CAPITAL PUNISHMENT OF ANIMALS, ETC.

NEW YORK
D. APPLETON AND COMPANY
1898

TO MY WIFE,
ELIZABETH E. EVANS.

CONTENTS.

CHAPTER	PAGE
INTRODUCTION: ANIMAL PSYCHOLOGY AS THE FOUNDATION OF ANIMAL'S RIGHTS IN THE HISTORICAL EVOLUTION OF ETHICS	1

EVOLUTIONAL ETHICS.

I.—THE ETHICS OF TRIBAL SOCIETY	19
II.—RELIGIOUS BELIEF AS A BASIS OF MORAL OBLIGATION	53
III.—ETHICAL RELATIONS OF MAN TO BEAST	82
IV.—METEMPSYCHOSIS	105

ANIMAL PSYCHOLOGY.

V.—MIND IN MAN AND BRUTE	165
VI.—PROGRESS AND PERFECTIBILITY IN THE LOWER ANIMALS	197
VII.—IDEATION IN ANIMALS AND MEN	222
VIII.—SPEECH AS A BARRIER BETWEEN MAN AND BEAST	270
IX.—THE ÆSTHETIC SENSE AND RELIGIOUS SENTIMENT IN ANIMALS	333
BIBLIOGRAPHY	359
INDEX	369

EVOLUTIONAL ETHICS.

INTRODUCTION.

ANIMAL PSYCHOLOGY AS THE FOUNDATION OF ANIMALS' RIGHTS IN THE HISTORICAL EVOLUTION OF ETHICS.

Recent enlargement of mental science. Close connection between evolutional ethics and animal psychology. Modern survivals of mediæval metaphysics and anthropocentric ethics. "Zoophily." Personification of inanimate objects by primitive peoples. Example from the Kalewala. Observation of animals by hunters and herdsmen in early society. Superstitious fear of animals and the rise of zoölatry. Survivals of animal worship in the cults of civilized races. Human appreciation of the lower animals as the result of their domestication. Their position as members of the tribe or family. Their worth recognised by primitive legislation. The dog in the Avesta. Zarathustra's care for cattle. Buddha's precepts in respect to animal life. The doctrine of evolution taught by Greek philosophers. The Ionic school of naturalists. Aristotle and Theophrastus. Greek speculation from Thales to Proclus. Celsus and Origen. Advanced views of Nemesius. His superiority to St. Augustine. Thomas Aquinas and the scholiasts. Beasts as types and symbols of spiritual truths. Their equality with man before the law. The principle of animals' rights asserted by evolutionists and generally opposed by theologians. Lotze's theory of soul and body. Psychical faculties as affected by the physical organism. Their coetaneous development and peculiar interdependence in the pithecoid stage of man's evolution. The starting point of humanity. General intelligence in the simplest organisms. Observations of Darwin and Romanes. Growth of instincts analogous to formation of habits. The measure of man's duty to the lower animals determined by the degree of their mental development.

THERE are scarcely any topics which excite such general interest, and are so frequently discussed nowadays, as the origin and evolution of ethical conceptions as revealed in the history of civilization, and the growth and development, the outward manifestations and essential qualities of mind in the lower animals, to the study of which the most recent researches in comparative philology, biology, psychology, and kindred branches of natural and mental science have given a fresh impulse and new direction, and opened up a broader and clearer field of view.

The intimate connection between evolutional ethics and animal psychology must be apparent to all who carefully consider the influence necessarily exerted by a proper appreciation of animal intelligence upon the recognition of man's moral relations and obligations to the creatures with whom he is so closely associated, and who are so largely subject to his dominion. The main argument urged by mediæval and modern scholiasts against the doctrine of the rights of animals is based upon the assumption that they are utterly devoid of those psychical powers which constitute personality even in the most restricted sense of this term. "Brute beasts," says the Rev. Joseph Rickaby, an English Jesuit and author of a work on moral philosophy, "not having understanding, and, therefore, not being persons, can not have rights. The conclusion is clear. They are not autocentric. They are of the number of things which are another's." He infers from these premises that "we have no duties of any kind to the lower animals, as neither to stocks nor stones"; "not of justice . . . and not of religion . . . not of fidelity . . . no duties of charity." * Father Rickaby and the Rev. Prof. Tyrrell represent a large class of dogmatic divines and belated schoolmen, who postulate an absolute and abysmal chasm between man and all other sentient organisms, and found upon this gratuitous assumption a narrow system of anthropocentric ethics at variance alike with the deductions of modern science and the finer feelings of humanity. In order to meet on their own ground these followers of St. Thomas Aquinas, the "angelic doctor," whose metaphysical quillets and *quodlibets* received the sanction of the Council of Trent and still rank as quasi-articles of faith in the Catholic Church, it is only necessary to show that the supposed chasm has no real existence as a fixed, final, and impassable barrier, and in the light of modern anthropological and psychological research has resolved itself into a wavering, indeterminable, and almost evanescent line of demarcation. As Miss Cobbe has very pertinently remarked: "The whole subject of our moral relations to the lower animals is undoubtedly a most obscure and difficult one. . . . Some revision of the 'Person and Thing' philosophy is, however, the first thing to be achieved; some reconstruction of the metaphysical and ethical systems of bygone times in better accordance with our present anthropolgy and psycholo-

---

* Quoted by Frances Power Cobbe in The Ethics of Zoöphily, a paper published originally in the London Contemporary Review (November, 1895), in refutation of similar views expressed by the Rev. George Tyrrell, also a disciple of Loyola.

gy.... The elephant and the butterfly can not be boxed together nowadays, except in a child's Noah's Ark. A Fuegian who eats his grandmother and can barely count his fingers can not be pigeonholed a 'Person,' and at the same time Landseer's dog a 'Thing,' except in a mediæval mind, which has somehow survived preternaturally into the Darwinian period." (Ib., p. 10.)

In tracing the history of the evolution of ethics we find the recognition of mutual rights and duties confined at first to members of the same horde or tribe, then extended to worshippers of the same gods, and gradually enlarged so as to include every civilized nation, until at length all races of men are at least theoretically conceived as being united in a common bond of brotherhood and benevolent sympathy, which is now slowly expanding so as to comprise not only the higher species of animals, but also every sensitive embodiment of organic life.

But while the primitive man regarded all human beings who were not his kinsmen as his enemies, his classification of the lower animals in their relations to himself was by no means so simple. In the childhood of the race, as of the individual, the imagination easily spans the gulf that separates the animate from the inanimate, and attributes consciousness and personality even to lifeless and formless objects. A striking illustration of this tendency, as it survives in poetry, is the manner in which Lemminkäinen, in the Finnic epos Kalewala, accosts the roadways which seem to come to meet her as she goes in search of her lost son:

> "Roadways, ye whom God hath shapen,
> Have ye not my son beholden,

Nowhere seen the golden apple,
Him my darling staff of silver?"
Prudently they gave her answer,
Thus to her replied the Roadways:
"For thy son we can not plague us,
We have sorrows, too, a many,
Since our own lot is a hard one
And our fortune is but evil,
By dogs' feet to be run over,
By the wheel-tire to be wounded,
And by heavy heels down-trampled."

The same naïve and vigorous fancy that could thus transform an *ensemble* of dust and clods into a living, thinking, and speaking entity would be still less cognizant of the spiritual disparity between man and beast, and would scarcely feel the absence of the "missing link," which modern anthropologists are making such strenuous efforts to discover. The grazing of flocks and herds, or the exciting perils of the chase, would lead to a close observation of the habits and peculiarities of different animals and give rise to strange conjectures and theories concerning their relationship to the human race, which in general qualities they so strongly resemble, and in special senses, such as sharpness of sight, keenness of scent, quickness of hearing, and swiftness of foot, they so far excel. The perception of these manifold capacities would suggest and enforce the recognition of an analogue of the soul underlying and controlling this complex of thoughts, feelings, impulses, and passions. Metaphysics had not yet woven its intricate raddle hedge of verbal definitions round the provinces of reason and instinct; the boundaries of the two spiritual realms were not so fixed, nor the distinctions so radical but that transitions from one state to the other were accepted as possible and even ordinary occurrences. Hence the popular belief in werewolves and other metamorphoses of men into beasts and beasts into men, which prevails in the primitive history, and survives among the lower classes of all nations, and plays so prominent a part in fairy tales and folklore, and forms the basis of the wonderful doctrine of metempsychosis.

Hence, too, arose a vague superstitious fear of the lower animals, not merely on account of their superior physical strength and natural ferocity, but also as embodiments of mysterious powers, and especially as reincarnations of deceased chieftains and warriors. This feeling is the source of totemism and the worship of deified ancestors in the forms of beasts and birds and even reptiles, which is probably the basis of all zoölatry. Survivals of this primitive cult are found in the mythologies of the most highly cultivated peoples, as, for example, in the eagle of Jupiter, the owl of Minerva, and the serpent of Æsculapius, where the animal, that was originally the real object of adoration, has become, in the evolution of religious ideas, simply the emblem of an anthropomorphic deity. Even Christianity, with all its spiritual aims and aspirations, shows distinct vestiges of zoölatrous worship in the conception of the Holy Spirit as a dove, of Christ as a lamb, of Satan as a dragon or a serpent, in the symbolism of the fish and lion, in the monsters of the Apocalypse, and the attributes of the evangelists borrowed from the vision of the prophet Ezekiel.

In this connection, however, our chief concern is not in the psychological explanation and historical evolution of zoölatry, but in its ethical influence as affecting man's treatment of the lower animals. The law of enmity is older and more universal than that of friendship. The earliest and strongest emotion in the breast of the savage is that of hatred and hostility to other men, as well as to all beasts of the field and of the forest. Indeed he makes no moral distinction between them, but regards them indiscriminately as foes, whom it is his imperative duty to destroy. If he recognises their superiority, he tries to flee from them or seeks to avert their wrath and win their favour by reverential submission and propitiation. In no case are they to him objects of affection; if he flatters them, it is not fondness but fear that is the motive of his conduct. The element of love does not enter into the religion of the primitive man, who adores and appeases by offerings and adulation only the beings he dreads.

The first feeling of genuine human sympathy with the lower animals grew out of their subjection and domestication, whereby they, like captives of war, were recognised as members of the family or tribe with which they were united by ties, not of actual affinity, but of adoption and common interest. They were reared and cherished because they contributed to the comfort and general welfare of the community, and this association during successive generations gradually led to the growth of permanent and traditional sentiments of kindness and benevolence toward them, and a natural desire to promote their happiness. The Sanskrit word for cattle (*paśu*) signified a creature "bound" to service, whether men, kine, horses, goats, or sheep, and the Roman *familia* included both domestic animals and slaves. The transition from the life of hunters to that of herdsmen, and finally from these nomadic stages to that of sedentary tillers of the soil, resulted in a more intimate knowledge and higher appreciation of the lower animals and a clearer conception of their mental and moral qualities. Man began to discover in them not only a remarkable capacity to understand him, but also a readiness and eagerness to execute his commands. This was especially true of his most faithful friend and constant companion, the dog, for whose proper nurture, protection, and kind treatment the sacred books of the ancient Persians contain the strictest injunctions with the severest penalties for their violation. These prescriptions, as well as those enjoining considerate care and compassion for cattle of every kind, although proclaimed as a revelation of the Good Mind (Vohu-manô) and embodied by Zarathustra in the Iranian religion, in order to invest them with supreme authority, were really based upon a perception of the intrinsic worth of the creatures themselves and their usefulness to man. This is evident from the distinction made between beneficent and baneful creatures, the latter being products and agents of the Evil Mind (Akem-manô emanating from the Hurtful Spirit Angrô-mainyush), which it is the sacred duty of the worshippers of the Living God (Ahuramazda, the personification of the Bountiful Spirit Spentô-mainyush) to exterminate. This dualism of god and devil is practically applied in the story of creation as recorded in the first fargard of the Vendidad, and furnishes the foundation of the most reasonable and equitable system of animal ethics developed by any Oriental people. Buddha forbade his followers to kill any animal whatsoever, and this absolute prohibition in the countries in which Buddhism prevails and ravenous beasts and poisonous reptiles abound, if conscientiously observed, would necessarily prove highly detrimental to the human inhabitants. But religious precepts, arbitrarily imposed, do not always suffice to curb the brutal instincts of the natural man, and the torture of animals, unwittingly through ignorance or wilfully through malice, is not unknown even in Buddhistic lands. In this, as in every department of ethics, the conduct of the individual depends upon the degree of his mental enlightenment and moral development, and is influenced by the religious creed he happens to profess only so far as the latter may incidentally modify his personal character. As a rule, its effect in restraining inborn propensities is very slight, especially when the religion is handed down from generation to generation as a sacred heirloom of the race and the performance of the duties it inculcates becomes perfunctory.

The metaphysical principle underlying this tender regard for all sentient organisms taught by Brahmans and Buddhists is the coessentiality of men and animals, from which the doctrine of metempsychosis is logically deduced. Many of the early Greek philosophers entertained the same theory, which was first fully developed by the Ionic school of naturalists and physiologists, one of whom, Anaximander, held the idea of evolution and even asserted the descent of man from the lower animals. It formed also the cosmo-theological basis of a system of animal ethics, most clearly and completely formulated, perhaps, in the writings of Aristotle's celebrated pupil Theophrastus. In fact, it pervades all Greek speculation for more than ten centuries, from Thales to Proclus, and is strongly emphasized by Plutarch, Plotinus, and Porphyrius, and other representatives of Neoplatonism and Neopythagoreanism, who made a practical application of it in urging abstinence from the use of flesh as an article of food. Indeed, this psychical homogeneity was so generally accepted by leading thinkers in the first and second centuries of our era as an unquestionable and quite axiomatic truth that the eclectic philosopher Celsus did not hesitate to adduce the denial of it as one of his most serious charges against Christianity. In replying to this acute and subtile, though rather superficial pagan polemic, Origen admits the correctness of the accusation, but is not at all disturbed by it; on the contrary, he maintains that the anthropocentric standpoint of Christianity is impregnable. All things, he declares, including animals, were created for man; the harmless ones being designed to be subjected to his will in order that they may minister to his convenience and comfort, while the hurtful ones contribute to the development of his thinking faculties and his sensibilities. How these latter effects are produced it is difficult to understand, unless it be by sharpening his wits in the struggle for existence against noxious creatures and by cultivating at the same time his patience and powers of endurance. So far as the animals themselves are concerned, Origen affirms that they have neither understanding nor will, but are 'mere mechanisms' skilfully constructed and kept in operation by the hand of God working through "all-mother Nature." This was the theory held by nearly all the Fathers of the Church and early Christian theologians, about the only notable exception being Nemesius, who was Bishop of Emesa in Syria during the latter half of the fourth century, and who seems to have believed with Ralph Waldo Emerson that

> A subtle chain of countless rings
> The next unto the farthest brings;
> . . . . .
> And, striving to be man, the worm
> Mounts through all the spires of form.

In other words, this remarkably clear-minded and sharp-sighted ecclesiastic, in his work on The Nature of Man (περὶ φύσεως ἀνθρώπου), appears to have discovered the principle of organic evolution fifteen centuries before Darwin made it the keystone of modern science, just as he anticipated Harvey by nearly thirteen centuries in describing the action of the heart and the circulation of the blood, and left on record some striking observations as regards the functions of the liver and the bile. He also maintained, in opposition to the current superstition of his day, that insanity is due to brain disease and not to demoniacal possession.

Nemesius, however, was endowed with a degree of insight and intelligence rare among his contemporaries and seldom shown even by the most enlightened of his coreligionists, among whom St. Augustine holds the first rank, not owing to superior learning, but on account of his uncommon intellectual acuteness, winning personality, and fiery zeal. As the chief exponent of the doctrine of predestination, the Bishop of Hippo robbed man of free agency and rendered him the wretched victim of divine decrees; but this affected only his relation to God and his eternal destiny, and did not diminish his dominion over "every living thing that moveth upon the earth"; an authority which patristic theologians, and especially mediæval scholiasts, with Thomas Aquinas at their head, claimed to be absolute and unrestrained by any recognition of rights or even sense of moral obligation on the part of man, except such humaneness and general benevolence as might spring from the vague and variable conception of his own worthiness.

Christian theologians and exegetists began also at a very early period to use the real or fabulous characteristics of animals for the illustration and enforcement of religious dogmas and moral duties. In this way it was possible to reconcile the existence of ravenous beasts and venomous reptiles with the omnipotence and beneficence of the Creator and Ruler of the world, since they were designed to serve as types and symbols of spiritual truths, and therefore held an important place in the system of redemption and consequently in the economy of the universe.* Still more interesting and inexplicable from a psychological point of view is the fact that not only rude tribes, but also highly civilized pagan and Christian nations have treated animals, otherwise deemed irrational, as though they were responsible for their actions, by placing them on a footing of equality with human beings as malefactors. According to the Mosaic law, an ox that gored a man or woman that they die was stoned, and this enactment has been often cited as a precedent by Christian tribunals in mediæval and even modern times in order to justify the execution of homicidal beasts. In Montenegro and other countries of eastern Europe horses, pigs, and horned cat-

---

* This subject has been fully treated in the author's Animal Symbolism in Ecclesiastical Architecture, published by William Heinemann in London and Henry Holt & Co. in New York.

tle have been tried for murder and condemned to death by criminal courts within the last half century. The highest ecclesiastical authorities have deigned to put the meanest vermin under ban, and not deemed it derogatory to their dignity to hold the terrors of excommunication over pernicious and disobedient locusts and slugs and vine-fretters.* This treatment of the lower animals would necessarily imply that their actions were regarded as justiciable, and that they stood in certain legal and therefore moral relations to mankind; for all law is ultimately based upon a more or less imperfect recognition of ethical principles, of which it aims to be the statutory expression.

But if animals may be rendered liable to judicial punishment for injuries done to man, one would naturally infer that they should also enjoy legal protection against human cruelty. It was a long time, however, before even the most enlightened nations reached this conclusion and began to form societies for its enforcement, and to give it practical efficiency by legislative enactments. It was in 1780 that Jeremy Bentham, in his Introduction to the Principles of Morals and Legislation and Principles of Penal Law, urged the duty of recognising and maintaining the rights of animals and asked, "Why should the law refuse its protection to any sensitive being? The time will come," he added, "when humanity will extend its mantle over everything which breathes. We have begun by attending to the condition of slaves; we shall finish by soften-

---

* For authentic accounts of such proceedings, see the author's work on The Criminal Prosecution and Capital Punishment of Animals, published by William Heinemann in London and Henry Holt & Co. in New York.

ing that of all animals which assist our labours or supply our wants." The ethical corollaries to Darwin's doctrine of the origin of species and to his theory of development through descent under the modifying influences of environment and natural selection have already passed these bounds of beneficence not only by demanding the mitigation of cruelty to slaves, but also by the abolition of slavery, and not only by inculcating the kind treatment of animals by individuals, but also by asserting the principle of animals' rights and the necessity of vindicating them by imposing judicial punishments for their violation. Penal laws having this object in view, but at first confined to the protection of neat cattle, were enacted in England as early as 1822; a little later they were made to include all domestic animals, and have been now greatly enlarged and adopted by nearly all civilized nations.

Only in countries like Spain, which are still governed by the antiquated metaphysical teachings and narrow moral theories of a mediæval hierarchy, has the *jus animalium* as yet found no place in codes of ethics or systems of jurisprudence. Even in Protestant lands, notwithstanding the distinctively humanitarian tendencies of the revival of learning and the reformation of religion in the sixteenth century, the clergy as a body has opposed every attempt to vindicate the rights of animals on scientific and zoöpsychological grounds as contrary to the teaching of Scripture. The German theologian Hettinger, in his Apology for Christianity, does not hesitate to denounce all such efforts to restrict the tyranny of man over the brute creation as the "ogling of materialists with beasts, which they seek to elevate merely for the purpose of degrading humanity." ⁵ No accusation could be more absurd; a correct conception of the origin and evolution of man and his kinship with the lower forms of life is essential to the proper appreciation of his dignity and destiny, and the full comprehension of his peculiar place in Nature. In this process of development it is impossible to separate psychical forces from physical factors, and to determine how far the faculties of the soul are dependent for their existence and exercise upon the structure of the body, inasmuch as we have no knowledge of the former except in organic association with the latter. According to the Neoherbartian philosopher Hermann Lotze, "all souls, considered as purely spiritual entities, are perfectly congenious or like-natured in perception, emotion, and will; but if the soul is incarnated in the body of an ape, it becomes an ape-soul, while in the body of a man it becomes a man-soul and mounts up to humanity. Souls are not different in themselves, but only in the degree of their development, and this depends upon the sum of the combined and varied excitations, which are conveyed to them. The more completely endowed and manifoldly equipped is the physical organism, the more perfect will be the soul, and this different grade of perfection constitutes the specific difference of the soul." This statement would seem to imply the creation and arbitrary distribution of souls, the exhibition of whose powers is dependent upon the physical conditions in which they chance to be placed. It would be more correct to assume that the soul gradually creates these conditions and produces a vehicle more highly organized, and therefore better suited to give line and scope to its full and free activity by diminishing the number and force of the predispositions and predeterminations, to which the nervous system of the lower animals is subjected. The immense intellectual disparity between a man of genius and a catarrhine ape is due to the accumulation of anatomical variations, so slight in their beginnings as to be hardly perceptible. This is especially true of the brain as the central organ of the nervous system, the increase of the surface of which through the multiplication of the folds and the deepening of the furrows marks the growth of intelligence and measures the increase of mental capacity. Other physical changes contribute to the same result: the assumption of an erect posture through the straightening of the legs and the formation of the firm, but elastic arch of the foot, thereby giving greater freedom of movement to the head and also to the hands as organs devoted exclusively to tact and prehension; the wider range and finer discrimination of the senses of sight, hearing, taste, and smell; and the superior flexibility of the glottis essential to articulate speech, all of which enable man to attain a more complete and exact knowledge—first, of his own body and secondly of the outer world—than it is possible for any lower animal to acquire.

But it would be wholly foreign to the purpose of this introduction to discuss the origin and nature of spiritual endowments, and the extent of their causal connection or correlation with physical characteristics; it suffices to show that the development of the former proceeds *pari pasu* with the development of the latter. In the primitive or pithecoid stage of humanity prehension was undoubtedly a more valuable and indispensable aid to comprehension than it is to-day; and the synonymy of a thin skin with a sensitive soul is the metaphorical survival of the actual workings of cause and effect in the earliest history of the race. But whatever may be the nature and extent of the interdependence between these physical and psychical elements, the development is everywhere a continuous one, with no break in the series of countless concatenations and marvellous adaptations of means to ends, by which the grand result is attained. The turning point in this endless and uninterrupted process of evolution, the point at which the beast ceases and the man begins, is where the soul is no longer the menial, but asserts its supremacy as the master of the body.

Researches in comparative psychology, taken in its widest sense as comprizing mental processes in the lower animals as well as in the lowest races of mankind, prove conclusively that even the simplest organisms are endowed with a certain degree of consciousness and so-called "general intelligence," as is evident from the analogy of their actions with those of human beings. Darwin affirms that "even the headless oyster seems to profit by experience," and Romanes maintains that the movements of an animalcule like the amœba indicate an intentional adaptation of means to ends; but the exercise of this power implies rationality as distinguished from that unconscious and involuntary impulse to action known as instinct. That the transformation of actions implying free intelligence into instinctive actions resulting in hereditary tendencies is constantly going on, and plays an important part even in the earliest stages of psychical evolution, there can be no question. In this respect the growth of instincts in the lower animals is analogous to the formation of habits in man.

This subject, however, has been so fully treated in the second part of the present volume that it is hardly necessary to make further reference to it here, except to point out its moral bearings. The measure of our duty towards lower organisms is determined by the degree of their mental development, or, as the German philosopher Krause has expressed it, "every creature endowed with a soul is also endowed with rights." The only firm foundation of animal ethics is animal psychology. It is through the portal of spiritual kinship, erected by modern evolutional science, that beasts and birds, "our elder brothers," as Herder calls them, enter into the temple of justice and enjoy the privilege of sanctuary against the wanton or unwitting cruelty hitherto authorized by the assumptions and usurpations of man.

It may be stated, in conclusion, that the contents of the present volume consist chiefly of articles which were originally printed in The Popular Science Monthly, The Atlantic Monthly, and The Unitarian Review, and which, after having been thoroughly revised and considerably expanded, are now offered to the public in a more convenient and more permanent form. A bibliography is appended, embracing the principal sources of information, and including also a number of works opposed to the author's views. The reader is thus aided in extending his studies, and by acquainting himself with the results of the latest researches enabled to form an independent judgment.

I.
EVOLUTIONAL ETHICS.

CHAPTER I.

THE ETHICS OF TRIBAL SOCIETY.

Limitations of the world of the primitive man. Relativity of geographical ideas. Survival of these conceptions in language. Ethnocentric ethics. The brotherhood of blood. Lactantius's theory of duty compared with the cosmopolitanism of Menander, Cicero, and Marcus Aurelius. General outlawry of aliens mitigated by the sacredness of hospitality. Spartan distrust and hatred of strangers. Tokens and tallies of friendship among Greeks and Romans. Supposititious kinship of tribal chiefs as a second stage in the growing conception of human brotherhood. Outcroppings of tribal ethics in the lower strata of civilized society. Clannish perversion of justice in Switzerland. Traces of this spirit in ancient French and German legislation. Old English alien laws a relic of savagery. Gradual recognition of the rights of foreigners in modern states. Insularism in British treaties of extradition. The tribe older than the family as shown by the social organization of anthropoid apes. Transition from nomadic to sedentary life. Influence of woman in effecting this change. Dwarfs and cripples as inventors. Why artificers in mythology are lame. Remarks of Mr. Maine on the supersession of tribal by territorial sovereignty. The Indo-Aryan as a "nigger." Weakness of race feeling in the United States. Strongest manifestations of it in the least cultivated portions of the country, toward the negroes in the South and Chinese in the West. The right of voluntary expatriation. Appeals to ethnic an-

20   EVOLUTIONAL ETHICS.

tipathies for political purposes: Latin Union, Panslavism, Panteutonism, and Anti-Semitism. Marriage of kin among the ancient Persians and Hebrews. Long survival of it as the sacred privilege of priests and kings.

THE world of the primitive man was bounded by the circle of his vision. He regarded the horizon as a fixed line which separated the earth from the sky, and which it would be possible for him to reach by going far enough. He did not deem it less real because it unfortunately always eluded his search, like the fabulous pot of gold which, according to popular superstition, lies buried at the point where the rainbow rests on the ground. In like manner the barbarian of to-day has no conception of the fact that the line of junction of earth and sky has no real existence, but is "all in his eye."

Indeed, it is but recently that man has learned to appreciate aright the wholly subjective character and significance of the terms north, south, east, and west as applied to places on the globe, and to recognise the relativity of all his geographical ideas, inasmuch as these are dependent for their accuracy and exactness upon the position of the speaker. It is one of the rare achievements of high culture, and has always been the prerogative of exceptionally thoughtful minds, to be able to distinguish between the apparent and the actual, to keep mental conceptions free from the influences of optical illusions, and not to be deceived by the surprises and sophistries of the senses.

An old English legend entitled The Lyfe of Adam, which has been preserved in a manuscript of the fourteenth century, relates how "Adam was made of oure lord god in the place that Jhesus was borne in, that is to seye in the cite of Bethleem, which is the myddel of the erthe." It then goes on to state that the first man was made out of dust taken from the four corners of the earth, which meet in Bethlehem, and that he was called by a name composed of the four principal planets: thus he was formed as a microcosm, the miniature counterpart and organic epitome of the universe, the synopsis and symbol of all created things.

There is a tendency in every savage tribe and isolated people to regard the portion of the earth which it happens to inhabit, and especially the spot which is the cradle of the race or around which its sacred traditions cluster, as not only the political and religious but also as the physical center of the world. Such were Jerusalem to the Jews and imperial and papal Rome, *urbs et orbis*, to the ancient Romans and mediæval Romanists; such has Benares been from time immemorial to multitudes of Hindus, and such is Mecca to-day to millions of Moslems. Before the discoveries of the Western hemisphere, made by Columbus and his compeers, not even the most enlightened peoples had any proper sense of their relations to the rest of mankind, either morally or geographically. International ethics and comities began with the growth of clearer and more correct ethnical notions, and have always kept pace with it. The knowledge of the rotundity of the earth gave a strong and permanent impulse in this direction, and has contributed not a little to the recognition of the equal rights of all races of mankind.

The language of every civilized nation contains curious survivals of the primitive conceptions which sprung out of what might be called the self-conceited and self-centered spirit of the savage. It is interesting to note how a single people, emerging from barbarism and taking the lead in civilization at an early period, imposes its forms of speech, and especially its geographical terms, upon after ages and upon remote races of men for whom they have really no meaning. We still speak of certain countries as the Levant and the Orient, the $Ἀνατολή$ of the Greeks, but these designations have no significance except for the dwellers on the shores of the Mediterranean, with whom they originated. So, too, Asia means etymologically the land of the rising sun and Europe the land of the setting sun, and these names expressed the actual position of the two continents in their relation to the Greeks. But to an American, and especially to a Californian, Europe is an Eastern and Asia a Western continent, and these strictly ethnocentric appellations would be wholly unsuitable and extremely confusing were it not for the fact that their etymology has become obscured and their primitive signification been forgotten, or is at least lost sight of and ignored, so that they are now mere arbitrary terms or distinguishing signs, with no suggestion of the geographical direction or situation of the regions to which they are applied, just as we speak of Chester, Edinburgh, Oxford, Berlin, or Munich without thinking of a Roman camp, King Edwin's castle, a ford for oxen, a frontier fortress, or a community of monks; and christen a child George, Albert, or Alexander without intending him to be a tiller of the soil, or wishing to imply that he is of noble birth, or will distinguish himself as a defender of men. All such proper names denote particular places or persons, but have wholly ceased to *connote*, as the scholastic philosophers were wont to say, the qualities or attributes which were at first associated with them and brought them into use.

The Chinese call their country the middle realm (*Chung-kuë*) or the flower of the middle (*Chang-hua*), thus characterizing it as the central and choicest portion of the earth, in distinction from the savage wastes inhabited by savage men outside of the Great Wall (*Wan-li-ch'ang-ch'ing*). The Jews looked upon themselves as the chosen people, set apart as *Yisrâêl*, or champions of the true God, and lumped all other tribes of men together as *goïm*, gentiles, poor pagan folks, who had no rights which a child of Abraham was bound to respect. The Greeks divided all mankind into two classes, Hellenes and barbarians; the latter were also called ἄγλωττοι—i. e., tongueless—because they did not speak Greek. Aristophanes applied the term βαρβαροι even to birds, on account of the inarticulateness and unintelligibleness of their chirpings and chatterings. It is from Greek usage that we have come to designate any corruption of our own language by the introduction of foreign or unfit words as a barbarism. The persistence of this primitive tribal conceit is shown by the fact that a people in many respects so cosmopolitan as the English can pronounce no severer censure and condemnation of the manners, customs, and opinions of other nations than to call them un-English, and really fancy that an indelible stigma attaches itself to this epithet. Not long since several British tourists in Italy actually protested against some foolish, perhaps, but otherwise harmless features of the Roman carnival, and demanded their suppression on the ground that they were "thoroughly un-English," thus virtually assuming that no amusements should be tolerated on the Tiber which were not customary on the Thames. It is due to the same feeling that the word "outlandish" has gradually grown obsolete in its original sense, and is now used exclusively as an expression of contempt. Slavonic (*slovene*) is derived from *slovo* (speech), and means people with articulate language; whereas the Slavic nations call the Germans *Němĭcĭ*, which signifies speechless, dumb, and therefore barbarian.

Geocentric astronomy and ethnocentric geography have been relegated long ago to that "limbo large and broad" which is the predestined receptacle of all exploded errors and illusions engendered by human vanity and ignorance; but from the bondage of ethnocentric ethics, manifesting itself in national prejudices and prepossessions, and often posing as a paragon of virtue in the guise of patriotism, even the most advanced and enlightened peoples have not yet fully emancipated themselves. The Hebrews thought they were doing the will of their tribal god (the personification of the tribal conscience) by borrowing jewels and fine raiment from their too-obliging Egyptian acquaintances and then running away with them. That this mean abuse of neighbourly confidence and civility was not a mere momentary freak of fraudulence or sudden succumbing to temptation, but the outcome of settled principles of morality and a general rule of policy, is evident from the approval with which it is recorded, as well as from the laws subsequently enacted, which permitted them to take usury of aliens and to sell murrain meat to the strangers in their gates.

This is the kind of ethics which finds expression in the legislation of all barbaric and semi-civilized races, from the Eskimos to the Hottentots. The Balantis of Africa punish with death a theft committed to the detriment of a tribesman, but encourage and reward thievery from other tribes. According to Cæsar's statement (De Bello Gallico, lib. vi, c. 23), the Germans did not deem it infamous to steal outside of the precincts of their own village, but rather advocated it as a means of keeping the young men of the community in training and rendering them vigilant and adroit. But we need not go to African kraals or American wigwams or primeval Teutonic forests for illustrations of this rule of conduct. Quite recently a Frenchman succeeded as *commis-voyageur* in swindling a number of German tradesmen out of large sums of money, and was applauded for his exploit by Parisian shopkeepers, who readily condoned his similar but slighter offences against themselves on account of the satisfaction they derived from the more serious injury done to their hereditary foes on the Rhine. This incident proves how easy it is for the primitive feeling of clanship, euphemistically styled patriotic sentiment, to put in abeyance all the acquisitions of culture and set the most elementary principles of honesty and morality at defiance. International conscience is a product of modern civilization, but it is still a plant of very feeble growth—a sickly shrub, whose fruits are easily blasted, and for the most part drop and decay before they ripen.

Sir Henry Sumner Maine, in his Lectures on the Early History of Institutions, has shown with admirable force and suggestiveness that rude and savage tribes uniformly regard consanguinity as the only basis of friendship and moral obligation and the sole cement of society. The original human horde was held together by the same tie of blood-relationship that produces and preserves the consciousness of unity in the animal herd or causes ants and bees to lead an orderly and mutually helpful life in swarms. In all these communities the outsider is looked upon as an outlaw; whoever is not a kinsman is a foe, and may be assailed, despoiled, enslaved, or slain with impunity. Indeed, it is considered not only a right but also an imperative duty to injure the alien by putting him to death or reducing him to servitude. The instinct of self-preservation asserts itself in this form with gregarious mammals and insects; and all primitive associations of men are founded upon this principle and cohere by force of this attraction.

A superstitious regard for blood pervades all early ideas and institutions of mankind. The ancient Hebrews were forbidden to eat the blood of a slaughtered animal, because the blood is the life; and the orthodox Israelite still clings to this notion and will not partake of butcher's meat that is not *gosh* or ceremonially clean—i. e., from which the blood has not been carefully drained off, although he knows that this process of ritual purification deprives the flesh of much of its succulence and nutritive value as food.

It is a widely diffused belief among aboriginal and lower races that the blood is the seat of the soul; hence blood-relationship is synonymous with soul-relationship. The child was also recognised as a blood-relation of the mother, but not of the father. Out of this conception of consanguinity arose the custom of descent in the female line, whereby the children of a man's sister became his heirs to the exclusion of his own offspring.

Curiously enough this notion is confirmed, to some extent, by modern science, which would ascribe to the female the function of conserving and transmitting the permanent qualities and typical characteristics of the race, whereas the influence of the male in propagation is variable, innovating, and revolutionary, and tends to produce deviations from the hereditary norm.

Cannibalism, too, as a tribal rite, originated in the belief that the soul resides in the blood, and that by drinking the blood of the bravest foeman their courage, cunning, and other distinctive and desirable traits may be acquired and thus serve to increase the fighting force and efficiency of the tribe.

Brotherhood was also created artificially or ceremonially by mingling a few drops of the blood of two persons in a cup of wine and drinking it. Each received into his veins a portion of the other's blood, and thus they became blood-related and were bound by the same mutual obligations as they would have been if the same mother had given them birth. The heroes of old German sagas are represented as drinking brotherhood in this manner; it is thus that Gunther and Siegfried swear inviolable friendship and fidelity in Wagner's Götterdämmerung; and German students, in the festive enthusiasm of a *Commers,* are fond of imitating their mythical forefathers in the solemn celebration of this mystic rite.

It is interesting to note the rhetorical and metaphorical survivals of this once strong conviction. In referring to political parties in France the Journal des Débats recently remarked: "It is not true that our nation consists of two nations—the heirs of the Emigration and those of the Revolution. This distinction no longer exists. The last vestiges of it have been obliterated on the battlefields, where all Frenchmen have mingled their blood. France is henceforth one and indivisible."

The noble sentiment expressed by the Greek comic poet Menander and handed down to us in the language of Terence, his Roman imitator, "I am a man, and regard nothing human as alien to me," was doubtless shared by many individual thinkers of antiquity, especially among the Greek Stoics and their Roman disciples. Cicero, who may be taken as one of the most eminent representatives of this ethical school, lays great stress upon "love of mankind" (*caritas generis humani*), in distinction from the love of kindred or countrymen. "A man," he says, "should seek to promote the welfare of every other man, whoever he may be, for the simple reason that he is a man"; and declares that this principle is the bond of universal society and the foundation of all law. He returns to this topic again and again, and never tires of enforcing this doctrine as fundamental in his treatises on duties (De Officiis), on the highest good and evil (De Finibus Bonorum et Malorum), and on laws (De Legibus). That he regarded this broad, cosmopolitan view as a new departure in ethics is evident from his remark that "he whom we now call a foreigner (*peregrinum*) was called an enemy (*hostis*) by our ancestors."

The distinguished Christian apologist Lucius Lactantius bases the duty of human kindness upon the hypothesis of human kinship, thus reviving and amplifying the old tribal notion which limits moral obligation to those who can claim a common progenitor. "For, if we all derive our origin from one man, whom

God created, we are plainly of one blood; and therefore it must be deemed the greatest wickedness to hate a man, even though he be guilty." He adds that "we are to put aside enmities and to soothe and allay the anger of those who are inimical to us by reminding them of their relationship. . . . On account of this bond of brotherhood God teaches us never to do evil, but always to do good." He also quotes a passage from the Epicurean Lucretius to the effect that "we are all sprung from a heavenly seed and have all of us the same father"; and draws from this statement the conclusion that "they who injure men are to be accounted as savage beasts."

Lactantius has been surnamed the Christian Cicero, but the fundamental principle of his ethics, as formulated in his Divine Institutions, is in its motive character and moral elevation far below the height attained four centuries earlier by his pagan prototype. The results of their teachings, practically applied, were equally cosmopolitan; inasmuch as Lactantius based his theory of duty on the Hebrew legend of the origin and descent of man, and thus enlarged his essentially tribal system of ethics so as to embrace the whole human race.

Marcus Aurelius defines his own ethical and humanitarian standpoint with his wonted epigrammatic terseness: "As an Antonine, my country is Rome; as a man, it is the world." Unfortunately, the liberal spirit of the philosopher, even when he happens to sit upon a throne, seldom exerts any direct and decisive influence in liberalizing the minds of the masses of mankind. Homer praises the kind and sympathetic heart of him who treats the stranger as a brother. But this fine sentiment does not change but rather confirms the fact that, as a rule, strangers were not thus treated in the Homeric age. As a general statement it remains true that in ancient times aliens had no legal rights whatsoever, and that international relations, so far as they existed at all, were relations of hostility.

But this outlawry *de jure* was mitigated *de facto* by investing the rite of hospitality with a certain sacredness. Such is still the case with all savage and semi-civilized tribes, as, for example, with the Bedouins, who hold the person of a guest inviolable, even though he may be their deadliest foe. This custom originated in the defenceless and helpless condition of the stranger, whose alienage placed him beyond the pale of law and the sphere of sympathy; it furnished a sort of compensation for the lack of all natural or conventional claims to protection, and thus supplied a temporary *modus vivendi*, without which intertribal intercourse would have been absolutely impossible.

We have an indication and illustration of this peculiarity of primitive society in the story of Cain, who, as a fratricide, was not only guilty of murder (a matter of comparatively small moment in the eyes of the aboriginal man), but also of treason against the tribe by violating the law of brotherhood fundamental to its constitution and essential to its existence; and when, by reason of this crime, he was driven out of the sheltering circle and sanctuary of his own kith and kin and became a fugitive and vagabond in the earth, his first feeling was the fear lest he should be slain by any stranger who might chance to meet him. The Lord is also represented as recognising the possibility of such a catastrophe, and as setting a mark upon him in order to avert it.

The stipulation contained in the Hebrew code, as well as in the code of other Eastern nations, which made it the duty of a man to wed his brother's widow, provided the first union was childless, and to raise up seed to the deceased, was only a modification of polyandry and differed from the conjugal relations still in vogue among the Thibetans in the fact that the possession of the same wife was successive instead of simultaneous. Both of these matrimonial customs are survivals of the earliest form of marriage, which was not individual, but tribal. We have a relic of this primitive kind of wedlock among the Californian Indians, who practised promiscuous sexual intercourse, so far as the members of the same tribe were concerned; the woman was regarded as faithless or adulterous only when she cohabited with a man belonging to another tribe.

The Greeks, with all their superior culture, never became as a people sufficiently enlightened to lay aside their deep distrust and depreciation of foreigners. Sparta was notoriously hostile to strangers (ἐχθρόξενος, or guest-hating), and how impossible it was for even a cultivated Athenian to look at the world at large from any but a strictly Hellenic point of view is curiously and comically illustrated in the drama in which Æschylus glorifies the battle of Salamis, where the Persians are made to speak of themselves as barbarians balked of their purpose, and to describe their lamentations over their defeat as dismal barbaric wailings.

It is a somewhat surprising and quite significant concession to Greek arrogance that Plautus should use the phrase *vortere barbare* in the sense of turning or translating into Latin. It is possible, however, that he may have borrowed this phrase from Philemon and other Greek playwrights, whose comedies he imitated with more or less freedom, but always with a touch of native genius. Still, we know that the Romans were uniformly called barbarians, and seem to have recognised the correctness of this appellation down to the age of Augustus, when the term began to be applied chiefly, if not exclusively, to the Germans. As our earliest information concerning the Germanic peoples was derived from Greek and Roman sources, we have been misled by the use of this depreciatory designation to think of them as wild and lawless hordes, and to form a wholly false conception of the grade and quality of their civilization.

When individuals of different race or nationality formed friendships they were wont to confirm the pact by an exchange of tokens, which remained as heirlooms in their respective families, and were prized by their descendants as pledges of mutually kind and hospitable treatment. The duty of helpfulness was, in such cases, quite as imperative as is the vow of *vendetta*, which passes as a precious inheritance of hatred from Corsican father to son. These tokens were called by the Greeks σύμβολα, and by the Romans *tesseræ hospitales*, and, although they were eventually superseded by better and more comprehensive methods and ended by playing only the frivolous part of a sentimental pastime in social life, like the modern philopena, they had originally a more serious purpose and were of no small importance as means of promoting intertribal intercourse and thus encouraging trade and leading to the establishment of commercial treaties.

Another step toward the realization of the conception of human brotherhood was the custom established at a very early period whereby chiefs of tribes came to address each other as kinsmen and members of one family. This assumption of consanguinity, which originated in the desire of dynasties to strengthen their position and to perpetuate their power, naturally led to increase of friendly intercourse and to frequent intermarriages, so that they finally became in fact what they at first claimed to be by a polite and politic fiction. Traces of this usage are found in the oldest records of royalty. Among the treasures of the Berlin and British Museums are preserved two hundred and forty-one tablets of cuneiform inscriptions containing letters written to Amenophis III and Amenophis IV of Egypt by Burnaburiash, King of Babylonia, and Dushratta, King of Mesopotamia, which show that, at least sixteen centuries before the Christian era, "dear brother" was the ceremonial title of salutation which monarchs were wont to use in their epistolary correspondence. This feigning of a common lineage still survives among crowned heads, and the vilest plebian adventurer who, by force or fraud, gets himself proclaimed king or emperor is admitted to the select circle of sovereigns and greeted as "dear cousin."

Principles, once grown obsolete, are denounced as prejudices; religious beliefs, which have been supplanted by superior creeds, are scoffed at as superstitions; and dethroned deities haunt the imagination of their former worshipper as demons. In like manner, the lower classes of civilized communities correspond, in a measure, to the lower races, and reflect atavistically the ideas and passions of primitive man; and in periods of great social and political upheaval we are often rudely brought face to face with tumultuous masses of these strata of palæozoic humanity violently and unpleasantly thrown to the surface. It crops out in the English boor, who at the sight of a stranger is ever ready to "'eave 'arf a brick at 'im," and would deem the neglect of this duty a treasonable lack of local patriotism and loyalty to time-honored tradition; in the Cretan herdsman, who instinctively seizes his cudgel whenever a traveller in trousers passes by; and in the Egyptian fellah, who teaches his children to spit at every man with a hat on and cry out: "*Yâ nasrânîy! Yâ khinzîr!* O you Nazarene! O you pig!"

The publican, in some parts of southern Italy, is still disposed to reckon with the foreigner as a foe, a forlorn vagabond, whom it is his native-born privilege to spoil. The blood of his ancestor, the brigand, courses in his veins, and his first impulse is to plunder the wayfarer. Prudence and the police may curb this progenital, predatorial proclivity; but the self-restraint always costs an effort, and, as a compromise with his instinctive feelings, instead of relieving the guest of his purse by force, he robs him of an undue portion of its contents by adding two or three hundred per cent to the usual price of fare and lodgment.

In many cantons of Switzerland, and especially in the Bernese highlands, we have the spectacle of a whole people apparently born and bred to consider mountain passes, romantic valleys, glaciers, and waterfalls as so many traps for curious and unwary tourists, and to prize sublime scenery merely as a ready-made snare to catch coots, dupes, gulls, boobies, and other varieties of too confiding summer birds of passage, which the categorizing mind of the German has reduced to two essentially distinct but closely connected classes, *Bergfexen* and *Sommerfrischler*.

This clannish spirit even invades and desecrates the courts of justice, and the Helvetian Themis is especially notorious for her propensity to blink the legal rights of the case and to tip the balance in favour of her cantonal or federal compatriots as opposed to the stranger within her gates.

In France the *droit d'aubaine* or *jus albinagii* confiscated to the crown the property of all aliens who died within the limits of the realm, to the exclusion of the natural heirs, unless these happened to be the king's subjects. This barbarous law was abolished by a decree of the National Assembly on the 6th of August, 1790, but was re-enacted twelve years later and incorporated in the *Code Napoléon*, modified, however, by a clause making the testamentary capacity of aliens dependent upon reciprocity; in other words, it was stipulated that the will of a foreigner should be declared valid in France, provided the laws of the said foreigner's country placed on the same footing the will of a Frenchman deceased within its jurisdiction. On the 14th of July, 1819, the *droit d'aubaine* was finally abrogated throughout the entire kingdom, after having been already considerably mitigated and partially annulled by the municipal authorities of Lyons and other industrial and commercial cities, which found this relic of mediæval legislation a serious obstruction to foreign trade.

Akin to this system of right was the German *Wildfangsrecht* or *jus wildfangiatus*, also known as *jus kolbekerlii*, which, as the term implies, accorded to human beings the privilege which game laws guarantee to the quarry, namely, that of being legally hunted. *Kolbenrecht* is equivalent to club law. An old and often quoted proverb, *Kolbengericht und Faustrecht ward nie schlecht*—the law of the strong was never yet wrong—is the cynical expression of protesting submission to the inevitable, recognised as outrageous. It is the same bitter sarcasm that mocks at unjust and irresistible power in the popular saying, "Might makes right"; it is despair taking refuge and finding relief in ironical humour, which turns the first principles of ethics topsy-turvy.

Wildfangsrecht was originally applied to fugitive serfs and to strangers, but was soon extended to bastards and bachelors, gleemen and professional champions in ordeals by battle, all of whom lived more or less in a state of outlawry as to their persons and property, and could, under certain circumstances, be reduced to the condition of chattels. Foreigners who could prove the place of their nativity were subjected to a poll tax (*chevage*) for the protection vouchsafed to them by the reeve or *Vogt*, and were therefore called *Vogtleute*. In the Canton de Vaud and elsewhere in Switzerland this pollage is still levied as *permis d'établissement*, a lingering vestige of mediæval extortion which the most enlightened European governments have now abolished. Persons of unknown origin were treated as waifs (*épaves*), the mere flotson and waveson on the drifting tide of humanity, and were liable to be seized and envassaled by any petty lord on whose territory they chanced to strand. Perhaps a diligent study of these old laws might suggest to American legislators some drastic means of purging the country of tramps.

In "the good old time" in England any alien could be arrested and punished for the crimes and misdemeanors of other aliens, although having no complicity with them. They were all lumped together as a class, any individual of which was liable to be apprehended and held accountable for the debts incurred or for the offences committed by any other individual of the class.

The idea of justice implied by such a proceeding corresponds to that entertained by the aboriginal Australian or American, who, when his wife dies, feels himself in duty bound to kill the wife of some member of another tribe, and avenges an injury inflicted upon him by a white man by slaying the first white man he happens to meet. The loss or offence, whatever it may be, is tribal, and is satisfied with tribal expiation or retaliation.

A case of this kind occurred quite recently in Dakota. A Sioux Indian, on the death of his squaw, went forth from his lodge with his gun and shot a missionary who was passing by. The red man had no grudge against the white man as an individual; on the contrary, he was personally fond of his victim, from whom he had received many acts of kindness; but the vow of vengeance was as sacred as that made by Jephthah the Gileadite, and had to be as religiously kept.

The old English custom, just referred to as a survival of the earliest and crudest conception of tribal ethics, prevailed at least as late as the reign of Edward III—i. e., till about the middle of the fourteenth century; and long after this period it was exceedingly difficult to enact and almost impossible to enforce laws for the protection of foreigners, so deeply rooted and intense was the prejudice against them. Even far down into the eighteenth century they continued to be regarded with extreme suspicion, and were often subjected to gross indignities, independently of any personal qualities or any peculiar conduct on their part. The mere fact of their alienage sufficed to kindle against them the anger of the populace and turn the masses into an unruly mob. Quite recently a Frenchman and his wife, who were attending a theatre in London near the Strand, went to an eating house close by to take some refreshment during a pause in the play. Very soon they were attacked by several persons of the lower class and severely beaten until they were finally rescued by the police. The sole provocation to this sudden assault was that they spoke a foreign tongue. This is still the mental attitude of the cockney, and cockneyism is only a local form of philistinism by no means confined to the precincts of Bow Bells.

The laws of Venice, as expounded by Portia in the case of Shylock *vs.* Antonio, discriminated against aliens as opposed to citizens in a manner extremely fatal to the plaintiff and exceedingly characteristic of mediæval legislation.

Under the influence of the political panic caused by the excesses of the French Revolution, Lord Grenville succeeded, in 1793, in persuading the British Parliament to pass an alien bill, in which the spirit of feudalism reasserted itself; and since the abolition of this retrogressive law, which was effected chiefly through the enlightened energy of George Canning, the leaders of the Tory party have repeatedly endeavoured to re-enact it. In every age and every country landed aristocracies have always shown a marked tendency to narrowness, provincialism, and distrust in their international relations. Indeed, from time immemorial, agricultural communities have been excessively conservative in this respect and hostile to progress; whereas commercial states and cities, whose prosperity is in proportion to their cosmopolitanism and dependent upon it, are naturally philallogeneal (to coin a word from the Greek of the Alexandrian patriarch Cyril, who unfortunately seldom exemplified in his conduct the virtue expressed by the epithet), or friendly to foreigners and easily accessible to influences from without.

Even in America, where all portions of the population are more mobile and undergo more rapid and radical changes than in other lands, the farmers are notoriously tenacious of old ideas and suspicious of reformatory movements of all kinds, following their traditions and clinging to their prejudices long after artisans and other handworkers of the manufacturing centers and large cities have cast aside these notions as obsolete and injurious.

All European governments appear to be periodically or epidemically affected with spasms of antipathy to aliens. France suffered from a particularly severe attack of this sort just before the Napoleonic *coup d'état*, and now betrays serious symptoms of a relapse, which it is to be hoped do not portend an imperial restoration. As a rule, such manifestations may be regarded as evidences of internal derangement, which is pretty sure to break out sooner or later in some violent disorder. Knownothingism in the United States was the symptom of such a crisis, although its indications were at that time only partially understood.

It is but recently, in fact, that civilized nations

have rid themselves of the most obnoxious relics of ethnocentric prejudice in their legislation—such, for example, as the *gabella hereditaria*, which discriminated against foreigners in matters of inheritance; and the *detractus personalis*, which virtually punished emigration by the imposition of a heavy fine. These vestiges of vassalage were removed from the statute-books of the German states in relation to each other by the acts of federation of 1815, and have been successively abolished between Germany and other countries by independent treaties.

The English law of extradition with other European powers still refuses to deliver up or to prosecute an Englishman who has committed a felony in a foreign land, unless the crime has been committed against one of his own countrymen. Some years ago a case of this kind occurred in Zurich, and still more recently in Munich. In the latter instance, one of the burglars, although residing in London, proved to be an American by birth, and was therefore handed over to the Bavarian police, and finally sentenced to ten years' imprisonment, while his English confederate in crime was set at liberty. Here we have, as the result of insularism, a survival of ethnocentric ethics in its crassest and most offensive form, such as one would expect to find only among a people still in the tribal stage of development.

In the volume already cited, Sir Henry Sumner Maine not only shows kinship to have been the original basis of society, but also indicates the process by which mankind may have gradually grown out of this primitive condition. The head of the family soon became through natural increase the head of a clan or tribe.

The patriarch possessed the authority and exercised the functions of a chieftain over his lineal and collateral descendants, who were known as his men and were called by his name. He was honoured and obeyed as their first man, *Fürst*, or prince, their stem-sire or king, an appellation which has nothing to do with personal "canning" or cunning, as Carlyle, in his excessive admiration of human force and faculty, would fain make us believe, but refers solely to race (*kuni*). The ruler was an ethnarch in the strictest sense of the term, and held his position by virtue of his primogenitureship or procreative seniority.

The correctness of this theory, so far as the genetic connection of the tribe with the family is concerned, may be questioned. Instead of the former being an aggregation or expansion of the latter, it is highly probable that the primitive tribe is older than the family and the product of promiscuous sexual relations, and that families originated in a subsequent process of domestic differentiation. Polyandry and the custom of tracing descent exclusively in the female line would seem to point in this direction. The institution of the family, even in its polygamous form, presupposes a certain ethical element, which can hardly be predicated of primeval barbarism.

So, too, the most prominent feature in the social organization of the anthropoid apes and in all simian communities is the troop or tribe under the leadership of the most powerful male. A band of orang-outangs is doubtless an association of blood-relations, but there is no recognition of patriarchal authority as such and no evidence of distinct divisions into families. The community is a gregarious group of individuals joined in affinity, but not yet separated into single pairs with clearly recognised and jealously defended conjugal rights; and sovereignty is simply the assertion of superior force, although this constitution of the simian tribe does not entirely exclude the existence and exercise of moral qualities in the mutual relations of its members.

It is, however, a matter of no moment for the further evolution of society, whether, at the beginning, the family expanded into the tribe or was gradually differentiated out of it. The fact remains that the tribe was held together by the cement of consanguinity, and that the authority of the tribal head was derived primarily from the respect and reverence due to him as common progenitor, aided, of course, by his ability to enforce his claims to rulership in case an ambitious and rebellious Absalom should be disposed to question them. So strong and persistent is this sentiment that, even now, the number of a man's noble ancestors is supposed to entitle him, by the grace of God, to sovereignty, or to confer upon him some exceptional privilege and power.

With the transition from a nomadic to a sedentary social state, an important change takes place. No sooner has a people acquired fixed habitations and established permanent settlements than there arises the idea of ownership in the soil, and the chief of the tribe becomes the lord of the land. He is no longer merely the head of an organized body of roving men, but he also claims and exercises jurisdiction over a more or less definitely circumscribed district or domain and over all persons dwelling within its borders. Tribal sovereignty or chieftainship is thus superseded by territorial sovereignty or dominion, and with this transformation the state, in the modern sense of the term, really begins.

At this early stage, however, proprietorship in land was not individual, but communal. It was the realization, to some extent, of the socialistic ideal of collective or governmental ownership of landed property, the return to which a modern school of reformers would fain persuade themselves and others to regard as a step in advance.

It is also interesting to note that this most important and epoch-making transition from pasturage to tillage was due to the initiative and activity of women. Everywhere in the growth of society women have been the first agriculturists. While the men were leading the life of hunters or herdsmen, with frequent episodes of pillage and predatory warfare, women began to cultivate the soil and to rear domestic fowls, to spin and to weave, and to develop, in a rude way, various kinds of industry. This is the condition in which we still find all savage and semi-civilized tribes. Herodotus (vol. vi) says of the Thracians, "They regard tillage as the most degrading and pillage as the most honourable occupation." The savage looks upon all forms of manual labour, and especially husbandry, as ignoble, and therefore leaves such work to his squaw.

At first, her efforts in this direction were quite ignored and often thwarted by the sudden removal of the tribe to another place before she could reap the fruits of her toil. The little patch of ground which she had planted was deemed of small account, compared with the pleasures and products of the chase, and was frequently abandoned without hesitation before the meager harvest was ripe. For this reason barley was the earliest grain cultivated, because it is the hardiest of all grains and matures soonest. It was a long time before the fields tilled by women became of sufficient importance, as supplying means of subsistence, to keep the tribe settled for a whole season in one spot, or even to induce them to return thither in the autumn and remain there until the crop was gathered. This semi-nomadism was the first step toward a sedentary life and the starting point of a higher civilization, and woman was the chief agent in its accomplishment, although unconscious of the immense change which her humble efforts were effecting.

For a similar reason the weakest male members of the tribe were the first artificers and mechanical inventors. Men who were crippled or otherwise incapable of waging war and following the chase, if they had not been left to perish at their birth, remained at home and made hunting implements and weapons of war for their more vigourous and valorous tribesmen, and thus acquired skill in handicraft, sharpened their wits, and developed their inventive faculties. In mythology, the gods of the smithy, Hephæstus, Vulcan, and Veland, are represented as lame, and the experts in ores and workers in metals are dwarfs, gnomes, and creatures of stunted growth. These physical peculiarities are not mere mythopœic whimseys and creations of the fancy, but correspond to real facts in the primitive history of the race, and point to the class of persons who were the earliest promoters of the arts.

The supersession of tribal by territorial sovereignty, although radical and permanent, was gradual and scarcely perceptible in its character, and did not begin to express itself in language till many centuries after the change had been fully accomplished. Mediæval and modern history furnish numerous illustrations of this process of social evolution and the manner of its operation. As Mr. Maine has remarked, there had been kings of England and of France long before John the Landless and Henry IV assumed respectively these official titles; although their predecessors had always been styled kings of the English and of the French. The Czar, who, while bearing sway as a territorial sovereign, preserves more than any other European ruler the peculiarities of a tribal chieftain, still calls himself Samodérshez, or Autocrat of all the Russians, and it was perfectly in keeping with the character and career of Napoleon I, as a *condottiere* on a colossal scale, that he took the title of "Emperor of the French." His interest was centered wholly in the army, which he loved and fostered in the same spirit that Tamerlane cherished his Mongolian hordes and Fra Diavolo his band of brigands. The King of Prussia bears the title of "German Emperor" (*Deutscher Kaiser*), not Emperor of Germany, since the latter would be inconsistent with the political existence and integrity of the other German states and a manifest usurpation of the rights and prerogatives (*Hoheitsrechte*) of the confederated princes and potentates. His imperial sovereignty is, therefore, essentially tribal; he is, so to speak, the chief of the German confederated monarchs, and exercises territorial sovereignty only as King of Prussia. There has been a long succession of Roman-German and German emperors, but never an Emperor of Germany.

A nomadic people, wandering from place to place,

is not associated in any sense with the soil; the tribe remains the same, but not the territory it occupies. With the beginning of agriculture and sedentariness this relation is reversed. The conception of a nation, nowadays, implies fixed or at least well-defined geographical boundaries. Changes may take place in the character of the inhabitants and in the constitution of the government as the result of emigration and revolution; individuals and families may disappear and be superseded by others of a different stock, but the nation remains, as it were, *adscripta glebæ* within certain territorial limits and is not destroyed by any admixture of foreign with native elements in the population. Mr. Maine states this point very clearly and concisely when he says: "England was once the country which Englishmen inhabited. Englishmen are now the people who inhabit England." An East Indian by blood may be an Englishman in the modern sense of the term as well as an Anglo-Saxon of purest lineage, however earnestly Lord Salisbury may deprecate the idea that a Hindu or any other "black man," even though he may be, like Dadabhoi Naoroji, a gentleman and a scholar, and the peer of the Tory premier himself in political wisdom and ability, should be sent to the British Parliament by an English constituency. It would seem, therefore, that, even at this late day, a man may be her British Majesty's first minister of state and yet entertain the notion, which prevailed in the days of Warren Hastings and still lingers among the subalterns of the colonial service, that an East Indian is a "nigger."

Nowhere is national feeling stronger and race feeling weaker than in the United States, where the negro, notwithstanding the prejudice growing out of his former condition of servitude, is as truly an American and as fully sensible of this fact as any scion of the Pilgrim fathers. It is unquestionable that the old Puritan stock is rapidly disappearing from New England, partly through natural extinction and partly through westward migration, and is being supplanted by Irish and Canadian French; but this circumstance does not blot New England from the map nor convert it into New Ireland or New France. On the contrary, the descendants of the Celtic immigrant are assimilated and transmuted by their environment and become New-Englanders. The consciousness of what might be called common territoriality tends not only to bind together and to blend diverse races into that " unity of a people " which constitutes a nation, but also to attenuate and to loosen the social and political unions, which are based upon common descent, and finally ruptures them altogether.

It is also in the United States that the antithesis to tribalism has found its strongest expression in legislation. In an act of Congress, approved July 27, 1868, the right of voluntary expatriation is declared to be " a natural and inherent right of all people, indispensable to the enjoyment of the rights of life, liberty, and the pursuit of happiness," and any denial or restriction of this right, or question of its validity, is affirmed to be " inconsistent with the fundamental principles of the republic." In fact, this enactment is only a reiteration and general application of the " self-evident truth " upon which the Declaration of Independence was based, and to the vindication of which by force of arms our Government owes its existence. It is the abrogation of the doctrine of personal and perpetual allegiance to the sovereign of one's native land, which is a survival of the notion, still prevailing among many savage nations, that the chieftain is the absolute owner of the members of his tribe and can dispose at will of their services, their property, and their lives. A strenuous effort to maintain this positon and to induce other powers to accept this principle has always been one of the chief features of the foreign policy of the United States, and in a few cases the Department of State has even carried the assertion of it to the verge of war. It was partially or conditionally acknowledged in the treaty of February 22, 1868, between the United States and the North German Confederation, and fully avowed in the so-called Burlingame treaty formed a few months later between the United States and China, the fifth article of which explicitly declares that both the sovereign powers "cordially recognise the inherent and inalienable right of man to change his home and allegiance."

In utter disregard of the principle involved in these treaty stipulations Congress has since then passed two acts practically denying the right of expatriation by refusing to accept its logical consequences—namely, the right of the individual thus expatriated to settle, labour, and become naturalized in the country to which he chooses to emigrate. The first of these acts was that of 1875 forbidding foreigners to enter the United States under contract to labour, and the second was that of 1882 excluding Chinese from the privilege of American citizenship. In both cases the abrogation of this "inherent and inalienable right of man" and "fundamental principle of the republic" was the result of demagogic pandering to the passions and prejudices of the lowest classes of the people, who still worship "the idols of the tribe" and show their faith by their works in burning negroes at the South and mobbing Mongolians in the far West. It is especially in remote and sparsely populated regions of the nominally civilized world that primitive barbarism survives and bears sway.*

The aborigines of British America, who can not regard human beings otherwise than from a tribal point of view, still speak of the English as King George's men; but the inhabitants of Canada consider themselves Canadians irrespectively of their ancestral origin, and the same readiness to sink the claims of lineage when they conflict with territorial interests manifests itself even in the more recent colonies of Australia and New Zealand. Geographical contiguity proves, in such cases, stronger than genealogical connections; the old proverb, that blood is thicker than water, does not hold true of oceans.

The appeals that have been made in recent times to ethnic antipathies and ethnic sympathies for the purposes of political propagandism or the promotion of personal ambition are anachronistic attempts to resuscitate the tribal spirit under new forms and on a larger scale by a perverse and pseudo-scientific application of the results of comparative philology to public affairs. The hobby of Napoleon III concerning the unity of the Latin nations, and the necessity of their closer confederation under the hegemony of France, was, like his Life of Cæsar, an act of historical self-justification, a desperate endeavour to explain his own *raison d'être*,

---

* Cf. An Abandoned Position in The Nation, vol. lvii, No. 1485, p. 443.

and thus set up a temporary prop to a rickety and rootless dynasty.

Panslavism may continue, for a time, to please the imagination and to fire the zeal of a people so peculiarly subjected, in many respects, to primitive social conditions and so powerfully swayed by primitive ideas as are the Russians; but Germany has long since outgrown the swaddling-clout of Panteutonism, and no ranting of anti-Semitic agitators and men of that ilk about *ur-deutsch* and *rein-deutsch* can permanently affect the public mind or elicit a favourable response in legislative enactments.

There is no cry so foolish or pernicious that it will not find a ringing echo in the empty brain-pan of some fanatic, no whimsey so silly and absurd that it will not be caught up and preached as a new gospel of universal redemption by a few pamphleteering demagogues or ill-balanced apostles of reform. Impecunious owners of poorly furnished and tenantless garrets are only too ready to let them to the first vagrant that knocks at the door, however seedy his appearance and doubtful his repute. Even the anti-Semitic crusade, so far as it has succeeded in getting a hearing and making any headway among sensible persons, has done so by appealing to the liberal spirit of the age and representing itself as a protest against the tribal exclusiveness of Judaism.

The constitution of the aboriginal tribe as a compact body of kinsmen, animated by feelings of hostility toward all other tribes, necessitated the intermarriage of blood-relations. If, on account of scarcity of females, or for any other reason, a man desired to wed a woman of another tribe, instead of wooing her as a friend, he waylaid her as a foe, stunned her with a blow of his war-club, and carried her off as booty rather than beauty to his camp, where she served him henceforth, not so much as his companion and helpmate as his slave and beast of burden.

Even after this tribal exclusiveness and isolation had ceased and a certain amount of amicable intertribal intercourse had grown up, it was still deemed more virtuous or, as we would say, more patriotic for a man to marry his own kin than to take his wife or wives from an alien people. The tribal religion also lent its special sanction to such nuptials. Survivals of this sentiment are found in the ancient customs and in the sacred Scriptures and traditions of many nations, especially in the Orient.

Thus, in the Avesta, a marriage of next of kin (*quaêtvadatha*) is declared to be particularly praiseworthy and well-pleasing to Ahuramazada, the Good Spirit (Visparad, iii, 18). This "kinship-union" is a prominent article of faith in the Mazdayasnian creed (Yasna, xiii, 28); and in the Book of Ardâ Vîrâf (ii, 1, 2) Vîrâf is said to have had seven sisters, who were to him as wives (*chigûn nêshman*), and this circumstance is adduced as evidence of his extraordinary piety. The connubial relations of this model of a religious man were both polygamous and incestuous.

Herodotus states (iii, 88) that Cambyses, the son and successor of Cyrus, was wedded to his own sister Atossa; and when, in the Hebrew story, Tamar rebukes Amnon for his guilty passion and tells him that "no such thing ought to be done in Israel," she refers solely to her brother's folly and wickedness in seeking a secret and illicit connection, and suggests that, if he will only speak to the king on the subject, there would be no obstacle to their union. That such marriages were common in the earlier history of the Jews is evident from the fact that Abram took to wife his half-sister Sarah, and this event is not recorded as an unusual occurrence.

Among the Persians this custom seems to have been confined, for the most part, to priests and kings, who constitute always and everywhere the two most conservative classes of society. Thus it came to be regarded as a mark of distinction or an enviable privilege, of which wealthy persons of inferior rank sometimes endeavoured to avail themselves; but there is no evidence that it remained, within historical times, a law for the entire nation or was generally practised by the people at large. The Magians continued to wive their sisters in conformity to ancient usage and holy tradition, for the same reason that stone knives and hatchets are used in sacrificial rites and fire for the altar is kindled by laboriously rubbing two sticks together long after these clumsy methods have been superseded in secular life by steel implements and lucifer matches.

CHAPTER II.

RELIGIOUS BELIEF AS A BASIS OF MORAL OBLIGATION.

The bond of blood superseded by the bond of belief. Theocentric attraction superior to ethnocentric attraction. The fiction of sacramental kinship in the Catholic Church. Religion as the cement of primitive society. Tribal religions nonproselytizing. Religious antagonisms in old Aryan society. Zarathustra's mission and creed. The worship of Ahuramazda and the holiness of agriculture. Inculcation of thrift and fruition by the Ahuryan religion. Condemnation of asceticism and celibacy. The begetting of sons as a means of salvation. The legend of Yima and the transition from pastoral to agricultural life. Antagonism between the good spirit and the evil mind. Modern examples of this enmity: Dards, Cossacks, Bedouins, and Mormons. Sinfulness of lending money on interest. Effects of this primitive notion in mediæval and modern times. Gradual growth of more enlightened views. Tribal spirit of Jewish burglars in Prussia. Uses of the Schabbesgoï. Brutality of the higher toward the lower races. Relapses into savagery through emigration. Moral restraint resulting from rapid international intercourse.

FOLLOWING the primitive period of tribal ethics comes a second stage of social and moral development, which Mr. Maine calls the supersession of the bond of blood by the bond of belief. Ethnocentric attraction gives way to what might be called theocentric attraction, and a broader and more spiritual sort of association is formed, having for its basis, not consanguinity, but conformity in religious conceptions. The god takes the place of the human progenitor of the tribe, or rather grows out of his deification in the evolution of ancestor worship, which is probably the oldest of cults.

Nevertheless, in this case, the fundamental principle of primitive society, which makes friendship coextensive with kinship, is not abrogated, but only enlarged in its application, causing those who worship the same deities or propitiate the same demons to enter into fraternal relations and call themselves brethren.

The canonical prohibition of marriage between persons connected merely by the artificial ties of a religious rite, such as sponsors and baptized infants, godfathers, godmothers, and godchildren, proves how intimately the idea of ritual relationship was associated with that of real relationship in the minds of those who established and perpetuated this institution. This fiction of sacramental kinship was at one time carried so far in the papal Church as to forbid the sponsor to be joined in wedlock even to the parent of a godchild. Cohabitation between a *patrinus* and a *matrina* was regarded as incest until the Council of Trent removed the ecclesiastical bar to such unions. The fact that they had assumed the position of spiritual parents to one infant prevented them from becoming the real and lawful parents of another infant. The importance attached to the name-day, which in most Catholic countries quite supplants the birthday as an anniversary, is also additional evidence of the vigour and vitality of primitive conceptions as embodied in ecclesiastical institutions.

Religion is, in fact, as Schelling observes, the strongest cement of primitive society, and the influence which contributes more than any other to the evolution and organization of the nation and state out of the tribe. Plutarch says: "Methinks a man should sooner find a city built in the air, without any ground to rest upon, than that any commonwealth altogether void of religion should be either first established or afterward preserved and maintained in that estate. For it is this that contains and holds together all human society and is its main prop and stay." Hegel expressed the same idea when he asserted that "the idea of God forms the general foundation of a people." Herbart calls attention to the pedagogical and disciplinary value of religion in the early stages of man's development, since it teaches him to subordinate present desires to future welfare, to look to the remote results of his conduct, and to sacrifice momentary pleasures here to permanent advantages hereafter.

But the ordinary experiences of life, especially in a cold climate, are quite as effective in inculcating thrift and enforcing the first elementary principle of domestic and political economy—that a man can not eat his pudding and keep it too. Stress of hunger emphasizes the necessity of laying up stores of provisions against time of need, and teaches foresight and forehand more directly and more forcibly than any hypothetical relation of man to the gods could do.

Originally the tie of religion must have been identical with the tie of relationship, and the brotherhood of belief coextensive with the brotherhood of blood, since all members of the same family or tribe would naturally adore the same domestic or tribal deities. Without this acceptance of the tribal theology and traditions by every individual of the tribe, the public peace would be constantly disturbed and the very existence of primitive society imperilled.

With the lapse of time and the increase of intelligence, however, vague wonder and ignorant worship would give place in more thoughtful minds to obstinate questionings, blank misgivings, and stubborn scepticisms, leading logically and inevitably to open schisms, and resulting in the formation of new communities of faith, crystallizing around the nucleus of a vital religious conviction. It was then proved, what all later history confirms, that spiritual affinities have a stronger cohesive attraction than natural affinities, and that, in every case of tension, the latter are sure to yield and be rent asunder.

Even the founder of Christianity, who professed to proclaim a gospel of peace on earth and good will to man, foresaw and did not hesitate to declare that this sundering of the closest consanguineous connections and division of families into hostile factions would be the necessary consequence of his teachings. He spoke of his doctrines as a sword destined to sever the nearest ties of natural affection and affinity, setting the son at variance against the father, and the daughter against the mother, and converting the members of a man's household into his bitterest foes.

The centre of cohesive attraction, which binds the new community so firmly together and so relentlessly ruptures all older associations, is the creed, or what is known in Christian theology as the symbol, the same term that, as we have already seen, was used by the Greeks to denote the token or pledge of hereditary hospitality and friendship between families, which furnished a basis for the formation of treaties of amity and commerce between tribes.

Strictly tribal religions never proselytize. Instead of seeking to share with alien tribes the favour and protection of their gods, they wish to monopolize whatever power and patronage may be derived from this source as a means of rendering themselves superior to their enemies. This was the case with the ancient Hebrews, who never thought of sending missionaries into other lands to make converts to Jehovah, but would have condemned such a procedure as treasonable. It is true that Jesus, in his denunciation of the Pharisees, declared that they "compass sea and land to make one proselyte"; but this reproof referred to their zeal as a political party in winning adherents among their own countrymen, in order to supplant the more liberal-minded and less rigidly ritualistic Sadducees in the Sanhedrin.

Jesus himself evidently never intended to break away from Judaism and to become the founder of a new religion. According to his own statement, he was "not sent but unto the lost sheep of the house of Israel." His mission was not to destroy, but to fulfil; not to abrogate, but to accomplish the law. He sought to give a spiritual interpretation to ancient precepts and injunctions; to revivify and rehabilitate the moral sentiment, hitherto dwarfed and deformed under the heavy burden of a perfunctory ceremonialism; and to enforce the commandments of God free from all incrustations of the traditions of men.

Curiously, and yet naturally enough, it was out of the very strictest sect of the Pharisees, so severely rebuked on account of their proselytic spirit, that the great proselyte Paul came—the man whose breadth of view and energy of purpose changed a local reformatory movement, which seemed to have been practically suppressed by the crucifixion, into a world-wide religion, by emancipating it from the fetters of Mosaic formalism, taking it out of the narrow *ghetto* of tribalism, and imparting to it a universal character. In this bold effort to turn apparent disaster into permanent victory, by breaking through the barriers of Judaism and preaching the gospel to the Gentiles, he met with the most determined opposition from the near kin and personal friends of Jesus, as well as from the principal disciples in Jerusalem.

To this process of development—by which Christianity, whose "field is the world," rose out of Judaism, the special cult of a privileged race—we have a parallel in the historical evolution of Buddhism, as a religion of pure humanity aspiring to universality, out of the narrow exclusiveness of Brahmanism with its rigorous politico-ethnological system of hereditary caste.

If, however, we go back to an earlier period, we meet with a most striking example of the workings of these conflicting forces in the disintegration and reconstruction of old Aryan society, thirty centuries ago, in the highlands of Bactria. The nature of this epoch-making movement, which took place as the result of Zarathustra's teachings and under his leadership, and the deep and enduring enmity it excited between people of the same blood, are perceptible in the solemn pledge or confession of faith by which the proselyte was received into the fellowship of the Iranian community.

This remarkable document, written in the ancient Gatha dialect, which is surmised to have been the vernacular of Zarathustra's native province and the mother-tongue of the prophet, begins with an abjuration of the ancestral deva worship and a vow of devotion to the glorious and munificent Ahuramazda, and then proceeds to a renunciation of all evil works, and especially of those deeds of violence peculiar to nomadic freebooters: "I choose the beneficent Armaiti (earth), the good. May she be mine! I detest all fraud and injury done to the spirit of the earth, and all damage and destruction to the homes of the Mazdayasnians. I permit the good spirits, which dwell on the earth in the form of good animals (such as sheep and kine), to roam undisturbed according to their pleasure. I praise, besides, all offerings and prayers to promote the growth of life. I will never do harm or hurt to the habitations of the Mazdayasnians, neither with my body nor with my soul. I forsake the devas, the wicked and malicious workers of iniquity, the most baneful, most malignant, and basest of beings. I forsake the devas and their like, the wizards and their allies, and all creatures whatsoever of such kind. I forsake them in thought, in word, and in deed. I forsake them hereby publicly, and declare that all their deceits and lies shall be put away." After further asseverations in the same strain, and after renouncing anew the devas, and entering into covenant with the waters, the woods, and the living spirit of Nature, and accepting the creed of the fire-priests, the diffusers of light and of truth, the convert concludes by avowing himself to be a disciple of Zarathustra, an adherent of the pure Ahuryan religion, and a member of the righteous brotherhood. Henceforth he is a sworn foe of the evil-doing, ancestral deities, and a zealous co-worker with Ahuramazda in promoting good thoughts, good words, and good deeds—*humata, hûkhta, huvarshta.*

With this proclamation of a purer religion the promulgation of a higher law of social life and a superior form of civilization was genetically connected—namely, the sacred duty of fostering and gladdening the spirit of the earth (personified as the goddess or angel Armaiti), by tilling the soil and making it fruitful. Husbandry is holiness to the Lord. In the third *fargard* of the Vendidâd this conception of agriculture as a sacred calling is particularly enlarged upon and enforced. The earth is there compared to a beautiful woman, who fails to fulfil her noblest functions so long as she remains virgin and barren. "He who cultivates barley cultivates righteousness, and extends the Mazdayasnian religion as much as though he resisted a thousand demons, made a thousand offerings, or recited a thousand prayers." Indeed, the best way to fight evil spirits is to redeem the waste places which they are supposed to inhabit. The spade and the plough are more effective than magic spells and incantations as means of exorcism. An old Avestan verse, which is quoted in inculcation and encouragement of tillage, and may have been sung by Iranian husbandmen as they sowed the seed and reaped the harvest, celebrates the influence and efficacy of their toil in discomfiting and driving out devils:

> The demons hiss when the barley's green,
> The demons moan at the thrashing's sound;
> The demons roar as the grist is ground,
> The demons flee when the flour is seen.

[These lines have also in the original a sort of rude rhyme or assonance peculiar to ancient poetry:

> Yadh yavô dayât âat daêva gís'en,
> Yadh s'udhus dayât âat daêva tus'en;
> Yadh pistro dayât âat daêva uruthen,
> Yadh gundô dayât âat daêva perethen.
> Vendidâd, iii, 105-108, Spiegel's ed].

If the Mazdayasnian religion, as revealed in the Avesta, illustrated in a remarkable manner the Benedictine maxim *laborare est orare*, it had no sympathy with the melancholy salutation *memento mori*, with which the Trappist greets the members of his silent brotherhood. As taught by the Iranian prophet and still practised by the modern Parsis, it is pre-eminently a religion of thrift, and enjoins as a sacred duty the honest accumulation and hearty enjoyment of wealth. Poverty and asceticism have no place in its list of virtues. Voluntary abstinence from the pleasurable things of the good creation is an act of base ingratitude and treason toward the bountiful giver of them. He who despises them is a contemner of Ahuramazda and an ally of the devas, and contributes thus far to the triumph of evil in the world. The righteous man should not dwell upon the idea of death, but banish it from his thoughts and earnestly strive after the realization of a fuller and richer life. It is the height of folly to suppose that mortifications of the flesh can further spiritual growth. Whatever fosters the health of the body favours the health of the soul; but the emaciation of the body impoverishes the soul. The notion which underlies what is known as "muscular Christianity" pervades the entire Avesta and finds a naïve and pithy expression in the following text of the Vendidâd, which the tiller of the soil is directed always to bear in mind and frequently to repeat:

> Who eateth not for naught hath strength,
> No strength for robust purity,
> No strength for robust husbandry,
> No strength for getting robust sons.

[Here, too, we have a bit of old poetry passed into a proverb. In the original the only trace of rhyme (and this we have preserved in the rendering) is the assonance of the second and third lines:

> Naêchis aquarentam tva,
> Nôit ughrâm ashyām,
> Nôit ughrâm vas'tryām,
> Nôit ughrâm putrôistêm.
>
> Vendidâd, iii, 112–115.

The editorial bracketing of the last line by Prof. Spiegel, as a possible interpolation, indicates an excess of critical suspicion, since this line not only fills out the verse, but also finishes up the thought, rounding and completing the expression of the sentiment with a climax.]

In another passage Ahuramazda declares: "Verily I say unto thee, O Spitama Zarathustra! the man who has a wife is far above him who begets no sons; he who has a household is far above him who has none; he who has children is far above the childless man; he who has riches is far above him who is destitute of them. And of two men, the one who fills himself with meat is filled with the good spirit (*vôhu manô*) much more than he who goes hungry; the latter is all but dead; the former is above him by the worth of a kid (*as'perena*), by the worth of a sheep, by the worth of an ox, by the worth of a man. [*As'perena*, usually rendered weight or coin, is derived from $a + s'par$, and means not walking or not grown, a young animal, a kid or a lamb. Cf. Sanskrit *sphar* or *sphur*, to expand or to swell.] Such a person can resist the onsets of As'tô-vîdhôtus (the demon of death); can resist the self-moving arrow; can resist the winter fiend, even though thinly clad; can resist and smite the wicked tyrant; can resist the assaults of the ungodly Ashemaogho (the destroyer of purity) who does not eat." (Vend. iv, 130–141.)

According to Herodotus (i, 136), the Persian king gave prizes to those of his subjects who had the greatest number of children. Vigorous procreation was one of the most effectual means of grace. It is stated in the Sad-dar that "to him who has no child, the Chinvad bridge (leading to paradise) shall be barred. The first question the angels who guard this narrow passage will ask him is whether he has left in this world a likeness of himself; if he answers in the negative, they will leave him standing at the head of the bridge, full of sorrow and despair." In the same work that contains this piece of eschatology it is also written: "There are those who strive to pass a day without eating and who abstain from meat; we, too, have our strivings and abstainings, namely, from evil thoughts, and evil words, and evil deeds. Other religions prescribe fasting from bread; ours enjoins fasting from sin."

The Brahmans maintained that the man who died without a son went to perdition, because there was no one to pay him the traditional family worship; hence the necessity of adopting a son in case he had none of his own. The Levitical law, as we have already seen, compelled a man to take the wife of a deceased brother, who died childless, and raise up seed to him. In the

Persian Rivâyats, or collections of traditions, similar matrimonial prescriptions are given. Thus, if a man over fifteen years of age dies childless and unmarried, his relations are to provide a maiden with a dowry and marry her to another man. Half of the children resulting from this union are to belong to the dead man and half of them to his proxy, the actual husband, and she herself is to be the dead man's wife in the next world. This kind of wife is called *satar*, "adopted." Again, if a widow, who has no children by her first husband, marries again, half of her children by the second husband are regarded as belonging to the first husband, and she also belongs to him in the future life; such a wife is called *chakar*, "serving." The first child of an only daughter belongs to her parents, if they have no sons, and they give her one third of their property in compensation. This kind of wife is called *yukan*, or "only child" wife. (Dr. E. W. West, Pahlavi Texts, in The Sacred Books of the East, vol. v, p. 143.) All these laws and customs show the vital importance attached to the possession of male offspring and to the preservation of an unbroken succession in the line of descent.

There are strong indications that the transition from pastoral to agricultural life in old Aryan society preceded the transformation of religious conceptions, and that the latter grew up gradually as a means of concentrating and more completely consolidating the former. In the second *fargard* of the Vendidâd a curious account is given of Yima, who lived before Zarathustra and is spoken of as a king rich in herds and a man of renown in Airyana-Vaêjô, the Eden of the race. It was this exalted personage whom Ahuramazda is said to have first chosen to be the promulgator of the true faith. But Yima, the son of Vîvanghañt (a name derived perhaps from *vangh*, to dwell or abide, and meaning settler or dweller in fixed habitations), excused himself, on the plea of unfitness for the prophetic office. He may have been, like Moses, a man of deeds rather than of words, "slow of speech and of a slow tongue." Then said Ahuramazda, "If thou wilt not be the bearer and herald of the faith, then shalt thou inclose my habitations and become the protector and preserver of my settlements." Thereupon he gave him a golden ploughshare and a goad decorated with gold as insignia of his royal office. [The word *s'ufra* I prefer to translate "ploughshare" rather than "sword" with Haug, or "lance" with Spiegel. It means literally a *cutting* instrument. In the Avesta, ploughing is called "cutting the cow"; and in the Vedic hymns the phrase "cut the cow" is equivalent to "make fertile the earth." "The soul of the cow" (*gêush urvâ*) means the spirit of the earth or the animating energy of Nature. In the Pahlavi translation of this passage *s'ufra* is rendered by *sûlâk-homand*, "having holes" or "sieve," and might therefore correspond to the Sanskrit *s'ûrpa*, "winnowing tray." The Pahlavi for ploughshare is *sûlâk*, and the close resemblance of this word to *sûlâk*, "hole," modern Persian *sûlâkh* and *sûrâkh*, may have led to a confusion and interchange of terms, both of which involve the idea of piercing or perforating.]

And Yima bore sway three hundred years; and the land "was filled with cattle, oxen, men, dogs, birds, and red blazing fires," until there was no more room for them therein. Then Yima went southward (literally, "toward the stars on the noonday path of the sun"), and, invoking the bounteous Armaiti, touched the earth with the golden ploughshare and pierced it with the goad; and, in obedience to his behest, the earth expanded and became one third larger than before. This process he repeated, according to the Zand, after six hundred years and again after nine hundred years, with a constantly increasing extension of the earth, which finally became about thrice its original size, and thus afforded ample space for men and kine.

It is not difficult to discover the meaning of this legend. It is the mythical statement of the effect of agriculture in practically enlarging the surface of the earth by increasing its capacity for supporting animal life, and thus rendering it possible for a greater number of persons to subsist on the products of the same area of soil. A tract of country which would furnish precarious food for a single hunter, or pasturage for a score of herdsmen, would, even under rude tillage, easily supply sustenance for a hundred husbandmen. Indeed, it has been estimated that one acre of arable land will bring forth as much food and consequently sustain as many inhabitants as two thousand acres of hunting ground.

In the fulness of time Yima was succeeded by the man who, like Aaron, could "speak well," and in the first Gâtha we find an address which Zarathustra delivered to his countrymen congregated around the sacred fire. It begins as follows: "I will now reveal to you who are here assembled the wise words of Mazda, the worship of Ahura, the hymns in praise of the good spirit, the sublime truth, which I see rising out of the sacred flames." He then appeals to them as the "offspring of renowned ancestors" to rouse their minds and give heed to his divine message: "To-day, O men and women, you should choose your creed."

After this brief exordium, he plunges at once into his subject and offers his solution of the old and ever-puzzling problem of good and evil, which he personifies as twin spirits, counter-workers in the creation of the world, each exercising its peculiar activity and contributing its characteristic element, and promoting respectively the happiness and the misery of mankind. It may also be safely asserted that, from a theistic point of view, no more logical and satisfactory solution of the difficulty has ever been presented. He earnestly exhorts his hearers to follow after the good and to eschew the evil. "Choose between these two spirits, for ye can not serve both." "Be pure and not vile." "Let us be such as help the life of the future." "Obey, therefore, the commandments which Mazda has proclaimed and enjoined upon mankind; for they are a snare and perdition to liars, but prosperity to the believer in the truth and the source of all bliss."

The whole aim of this discourse, of which these extracts suffice to indicate the drift, is to persuade his hearers to renounce or to confirm them in their renunciation of the old Aryan polytheism and worship of the devas, as we find it in the Vedas, and to adopt monotheism or the adoration of the one great and good but by no means omnipotent being, Ahuramazda. As a philosophical system, his doctrine was dualistic and recognised the existence of two original and independent principles in the universe; as a cult, it was monotheolatrous and worshipped only one of these powers.

It may be added that long before the close of the

Vedic period the Indo-Aryans had also begun to devote themselves to husbandry, although their chief wealth still consisted in herds. The burden of their hymns and prayers to the gods is for much cattle and a large family of vigorous sons. The foes which they now had mostly to contend with were the Dasyus or aborigines of India. The occasional mention of Aryan enemies may be partly reminiscences or records of an earlier time and partly references to intertribal warfares, of which there was evidently no lack. It must be borne in mind that all the Vedic hymns appear to have been composed in northern India, and principally in the region now kown as the Panjâb. In none of these poetical productions do we find any distinct remembrance of a trans-Himalayan origin or any definite allusion to a former residence outside of India. This circumstance proves that at the time of the supposed migration from the North the ancestors of the Indo-Aryans must have been rude barbarians, destitute not only of written records, but also of the ability to preserve and transmit from generation to generation traditions of great events in their own tribal or national history. The savage has a short memory for whatever lies beyond the sphere of his individual experience.

One of Zarathustra's chief injunctions was to "listen to the soul of the earth," and to "succour and foster the life of Nature." This is to be done by cultivating and fertilizing the soil; since the increase of its productivity augments the sum of vitality in the world and contributes to the ascendency of the *vohumanô* or good mind, synonymous with *vis vitalis* or living force, and aids in securing the supremacy of Ahuramazda. Instead of bowing down in servile fear before the phenomena of Nature, the Mazdayasnians are directed to revere and cherish her kindly and beneficent spirit, so that "the wilderness and the solitary place shall be glad for them, and the desert shall rejoice and blossom as the rose."

Angrô-Mainyush and his satellites, the devas, on the other hand, are constantly striving to resist and to thwart this purpose and to keep the earth in her native state of virginal wildness and ruggedness by investing her with the dread sanctities and superstitions of a crude polytheistic physiolatry, by assaulting and ravaging the cultivated settlements of the Ahuryan agriculturists, and by fomenting and fostering the spirit of primeval savagery, personified as *Akemmanô*, or the evil mind. In the sacred books and traditions of both factions, and more especially in those of the reformatory party, are frequent traces of this social rupture and religious schism, and of the deadly hostility naturally existing between nomadic hordes, that still adhere to a life of pasturage and pillage, and men of more advanced ideas, who dwell in fixed habitations (*gaêthas*) and devote themselves to husbandry.

I am well aware that M. James Darmesteter and other representatives of what might be called the meteorological school of Avestan scholars deny the historical reality of a religious schism of the kind here described, and would reduce Zarathustra and all the incidents of his life to a series of solar myths. It is, however, only on the theory of a religious schism that the fact that the deities of Brahmanism are the devils of Zoroastrianism, and *vice versa*, can be adequately explained. To assert that this antagonism is the result of an "accidental selection" of gods is no explanation at all. The religious history of mankind is not a record of casualties or mere chapter of accidents.

Besides, we have a modern example of a similar enmity growing out of the transition from nomadic to sedentary life in the mythology of the Dards, who are, perhaps, one of the oldest races and most primitive peoples of the East, and who believe in the existence of demons called *yatsh* (bad), which, like the Homeric Cyclops (the barbarous aborigines of the Sicilian coast), are of gigantic stature, and have only one eye, set in the middle of their forehead. These demons haunt the mountains and the wilderness, and are exceedingly hostile to agriculturists, whom they vex and harm in every possible manner, stealing and destroying the crops, and even carrying off the husbandmen to their gloomy caverns. In this scrap of mythology we have the survival of the old strife between barbarism and civilization, which began with man's first efforts to improve his condition.

The barbarian is, in fact, the most uncompromising incarnation and typical representative of conservatism; and it is the survival of the barbarian temper of mind that constantly hampers progress and hinders reform in modern times. His daily life is the dullest routine and would be unbearable, were it not the outcome and expression of the general rigidity and sterility of his intellect. He treads religiously in the footsteps of his forefathers, generation after generation, the whole mass moving on bodily and mentally in single file, as is the custom with savages. He is the stubborn foe of all innovations, and punishes as treason against the tribe every deviation from the beaten trail. Under such circumstances no social transformation can be effected without fierce battle and bloodshed. In the primitive history of mankind, as in the early physical history of the globe, great changes are uniformly the result of great convulsions.

It is not merely the love of booty that leads nomadic tribes to attack and lay waste the permanent settlements of husbandmen, but the instinct of self-preservation resisting the encroachments of a new form of social organization which imperils the old. For this reason hunters are hostile to herdsmen, and herdsmen to tillers of the soil; since pasturage diminishes the extent and value of hunting grounds, and agriculture diminishes the area of pasturage.

Mr. D. Mackenzie Wallace gives a striking illustration of this antagonism in the history of the Cossacks of the Don, who, so long as they lived by sheep-farming and marauding, prohibited agriculture under pain of death. This severe interdict of a peaceful pursuit originated, not as some have supposed in the desire to foster the warlike spirit of the people, but rather in a perception of the fact that "the man who ploughed up a bit of land infringed thereby on his neighbour's right of pasturage." By this act he became in a certain sense guilty of treason against pastoral society, the very foundations of which, the green sod, he broke up and destroyed with his ploughshare. He not only restricted and reduced the actual area of grazing, but also struck a blow at the life of a cattle-rearing community. The practical workings of this crude and clannish conception of patriotism are recorded, as Mr. Wallace observes, on the pages of Byzantine annalists and old Russian chroniclers, who describe the periodical havoc of farmsteads committed by the nomadic tribes which from time immemorial had roamed the vast plains north of the Black and Caspian Seas, razing the houses, ravaging the fields, and leaving the bodies of the husbandmen as food for vultures.

The roving Bedouins, dwellers in the desert, as their name implies, despise the cultivators of the soil and call them contemptuously *fellahin* (ploughers, boors); and their kinsmen the Anasis (*anâsî*, men) hover on the borders and levy blackmail on the villages of Syria. It is also significant for the persistency of this primitive point of view that the Arabic word for agriculture (*falâhat*), should also mean " fraudulent traffic," as though the permanent possession of a piece of land and the exclusive use or sale of the products of the soil were in themselves swindling operations.

These facts of to-day suffice to show the kind of opposition which Zarathustra had to face in his efforts to establish the Iranians in fixed settlements and to accustom them to the acquisition and proper utilization of landed property. In order to accomplish this purpose it was necessary to teach the holiness of husbandry and to invest seedtime and harvest with the sanctity of religion.

The Mormons, after their migration to Salt Lake, where the very existence of the community depended upon converting the desert into a garden, inaugurated the same policy, declaring through the mouth of their prophet that the human race could be redeemed and paradise regained only by means of tillage and making agriculture a sacred vocation and the pursuit of it a prominent part of their creed.

The priests of the old deva cult, the progenitors of the Brahmans, on the other hand, denounced Zarathustra as a schismatic and a renegade, a contemner of the gods and blasphemer, a scorner of ancient custom and subverter of social order. They therefore opposed the innovation and fought for the faith of their fathers with such clumsy weapons as they were most skilled in wielding, looting the homesteads, uprooting and trampling down the green blades of wheat and barley, which stood as representatives of the growing heresy, and, with a logic peculiar to theological zealots and ecclesiastical inquisitors in all ages, refuting the new doctrine and resisting the reformatory movement by greater energy and assiduity in the ancient and honourable calling of cattle-lifting.

As we have already seen, the duty of a man to shield and sustain a tribesman against an alien under all circumstances is imperative. Acts of extortion, treachery, or violence, which would be punished by death if committed against a member of the same tribe, are regarded as indifferent or laudable when the injured person is a foreigner. The same tendency to approve or to extenuate the bad conduct of "brethren" enters also more or less into the ethics of all communities or collective bodies which are held together by the bond of belief.

All people in a low state of civilization have a strong prejudice against lending money on interest, and look upon all such transactions as sinful. The same notion still prevails among the lower classes of civilized nations, whose superstitions are in most cases mere survivals of savage life. So strong is this feeling, inculcated and consecrated by religious teachings and traditions, that a certain stigma attaches to the money broker even in the minds of otherwise intelligent persons. "Many lend money on interest," says Cato, "but it is not honourable to do so. Our ancestors enacted in their laws that the thief should restore twofold, but the taker of interest fourfold, from which we see how much worse a usurer was thought to be than a thief."

In general, however, usury, like every other supposed crime, was regarded as wrong only when applied to kindred or tribesmen. The Jews were forbidden to "take a breed of barren metal" from those of their own faith, but might exact it from Gentiles. Curiously enough, in the middle ages this privilege was granted to the Jews, not in the spirit of favouritism, but as a necessity to sovereigns and to society and from feelings of utter scorn and contempt. As neither government nor trade could do without this vilely esteemed vocation, the Jews were selected to carry it on, because they were considered a vile people incapable alike of improvement or of deeper degradation. The state and the Church, which felt an interest in the spiritual welfare and safety of the Christian, were wholly indifferent to the future fate of the Jew. That sweet saint, Bernard of Clairvaux, surnamed the honey-flowing teacher (*doctor mellifluus*), urged the rulers of his day to tolerate the Jews, not because he hated persecution, but in order that Christians might not be constrained to imperil the salvation of their souls by the sin of usury. The Israelitic pariahs of mediæval society rendered the same service to Christian virtue that professional prostitutes do to female chastity. We have a striking illustration of this point of view in a decree issued in 1219, by the German emperor Frederick III, permitting the Jews to dwell in Nuremberg and to take a percentage for the use of money. Inasmuch as this business, he said in justification of his edict, is essential to the growth of commerce and the prosperity of the city, it will be a lesser evil and wrong for Jews to practise usury than for Christians, since the former are a stubborn and stiffnecked race, and, if they persist in their perversity, as they probably will do, are doomed to be damned anyhow.* We have a relic of this primitive prejudice in the efforts of modern governments to establish a fixed rate of interest for the use of money and to punish as usury any higher compensation for it. All such attempts have uniformly proved to be not only futile, but also productive of evil to both borrower and lender, and especially to the former; and as the result of more enlightened views of financial and economical science they are gradually sharing the fate of sumptuary laws and similar regulations and disappearing from the statute books. The value of money, like that of any other marketable commodity, can not be positively prescribed by legislative enactments, but must be determined by the natural law of supply and demand.

The Hebrew, on the other hand, heartily reciprocated the Christian's contumely, and could hardly conceal, under the prudent disguise of mock humility, his disdain for the upstart Nazarene. He not only deemed it a religious duty to cheat him in money matters, but thought it perfectly right to use him as an agent in base or criminal transactions which a good Israelite could not conscientiously perform.

This mental and moral attitude, which even the

---

* We have referred to this characteristic decree in a work entitled Animal Symbolism in Ecclesiastical Architecture (London: William Heinemann; New York: Henry Holt & Co., 1896, p. 293) for the purpose of illustrating another subject.

modern Hebrew still maintains, is strikingly exemplified by the following incident: Between 1820 and 1830 a band of burglars, numbering over one hundred persons and consisting entirely of Jews, made property so unsafe as to create a panic among the inhabitants of the Prussian provinces of Posen and Brandenburg. The chief of the band was a certain Loewenthal in Berlin, and all the members of it were extremely devout attendants of the synagogue and strict observers of every jot and tittle of the Levitical law. They never broke into the houses of Jews and never stole on the Sabbath, since such an act would be a desecration of the sacred "day of rest"; but, rather than let an exceptionably favourable opportunity escape, they sometimes employed a so-called *schabbesgoï* [*schabbesgoï* (Sabbath-Gentile) is a Jew-German term for the Christian attendant or servant who does for an Israelite on the Sabbath the things which his religion forbids him to do for himself] to commit the crime for them, and, if necessary, did not hesitate to have some one of their own number accompany him on his burglarious expedition a couple of thousand yards or so, the limits of a Sabbath day's journey. In case one of the band was suspected of any particular offence and arrested, the surest and speediest way of clearing himself was to prove an *alibi* by the testimony of two witnesses, as the law required. But the pious Hebrew regards perjury with peculiar abhorrence, and fears above all things to take a false oath. Shylock was eager to cut the heart out of his hated enemy, but he would not lay perjury upon his soul—no, not for Venice! The burglars kept, therefore, in their pay two Christians, who were as ready to forswear themselves as any Tammany Hall politician at the polls, and who made the requisite false oaths at fixed rates.

These examples serve to show the natural tendency of mankind to look upon compatriots and coreligionists from a different moral standpoint from that with which they regard persons who are not connected with them by such ties, and to whom they not only attribute a lower standard of right and wrong, but also act upon it as a rule of conduct in dealing with them.

Great dissimilarity in physical characteristics intensifies the ethical estrangement caused by differences of blood and of belief. The more any tribes of men deviate from ourselves in form and feature, the less we are inclined to think of them as endowed with the same powers and passions, the same kind of sympathy and sensibility as ourselves, or as entitled to the same rights that we possess. A people with black skin, woolly hair, flat noses, and countenances of a strongly prognathous character do not enlist our kindly feelings and awaken our affections in the same manner and degree as representatives of a fair-complexioned and finely featured type would do. The schemes of European governments and of private individuals and corporations for the exploration, partition, and colonization of Africa are based upon the assumption that the Africans themselves have no claim to the continent which they inhabit. The only African colony that has ever been founded on principles of common justice and with a full recognition of the rights of the natives is the Republic of Liberia, established more than sixty years ago under the auspices of the United States, and this was done solely for the sake of getting rid of an undesirable population of free negroes at home. All the other enterprises of this sort are morally and legally no better than buccaneering expeditions.

The ethical maxims which we are wont to accept as axiomatic in our mutual relations as civilized individuals and nations are too easily set aside as inconvenient and inapplicable to our dealings with the so-called lower races. The fatal facility with which under such circumstances enlightened Europeans of the nineteenth century may revert to primitive savagery as soon as the outward restraints of civilization are removed is seen in the early settlers of Australia, who did not scruple to shoot the defenceless and harmless aborigines as they would any game, and feed the carcasses to their hounds. The inoffensive and rather feeble-bodied Negritos were treated as beasts of venery, which could be hunted without danger and furnished plentiful supplies of dog's meat, costing the sportsman nothing, not even a pang of conscience, only the price of a cartridge. (Cf. Schaafhausen, in The Anthropological Review, London, 1869, p. 368.)

More recent and even more revolting exemplifications of this tendency to relapse into barbarism are the atrocities committed by Major Barttelot, and the conduct of Mr. Jameson, of Stanley's Emin-Relief Expedition, who purchased a young negro girl and gave her to a horde of cannibals in order to make sketches from life of the manner in which she was torn in pieces and devoured.

The atrocities still committed by the officials of the Belgian Government in Congo are a disgrace to a civilized people. Scores of natives have their hands cut off or are otherwise mutilated simply because they are unable to supply ivory and rubber enough to satisfy the insatiate greed of traffickers in those articles. Soldiers in the service of the State are permitted to eat the bodies of those who have fallen in battle, since human flesh, thus obtained, furnishes the cheapest rations for the army. As the result of this policy races, who were not cannibals when they first came in contact with white men, have gradually become so through intercourse with cannibal troops under the command of Belgian officers. Thus the increase of cannibalism on the Congo is due to the domination of a European sovereign acting as the representative of the European powers.

There are also instances on record of Englishmen, Dutchmen, and Frenchmen who in their warfare with Indians adopted from their savage foes the custom of scalping and torturing their captives. In fact, as Waitz has shown in his Anthropology (iii, 174), there is scarcely a vice of barbarous tribes which Europeans when removed from the restraints of civilization have not practised. In the South Sea islands they have in some cases become anthropophagous.

Here we are suddenly brought face to face with the depressing fact that men, who are heirs to ages of intellectual culture and armed with all the powers and possibilities of good and evil which modern science has put into their hands, yet relapse morally to the level of rude cave dwellers and contemporaries of the mammoth in making their superiority of mental endowment and material equipment minister to deeds and passions worthy of the lowest stage of barbarism.

All emigration to wild regions is, in a greater or less degree, atavistic in its effects, and, by loosening or removing the many leading strings of association by which the average man is kept in an upright position and a straightforward course, lets him fall back and retrograde, and thus tends to bring him nearer to his flint-chipping neolithic ancestor. It throws each individual upon his own ethical resources by releasing him from the constant though hardly conscious social pressure of an environment which is the resultant of long periods of human progress, and by which alone the masses of so-called civilized nations are prevented from relapsing into the original condition of the race.

Happily, however, such extreme cases of moral reversion as those of the early emigrants to Australia and the recent explorers of Africa are only sporadic, and the ubiquity of humane and enlightened public opinion arising from greater frequency and rapidity of international intercourse, and causing its immediate influence to be felt in the remotest and roughest border lands of savage and civilized life, will render them still rarer in the future. The telegraph and the telephone are making it daily more difficult and will eventually make it impossible for the most pushing pioneer wholly to lose communication with the advancing body of organized forces behind him, or to break away from the control of that community of impulses and purposes, and that consensus of moral ideas and perceptions, which we call public conscience. This influence is beginning to penetrate even the darkest regions of Central Africa and to protect the unknown barbaric tribes against the ravages of Arab slave traders and the arbitrary authority of European adventurers. Each nation that joins in this combined movement is doubtless seeking, first of all, to further its own commercial and colonial interests; but it suffices as an illustration of the prevailing spirit of the age that the basis on which they profess to unite is the broad principle of a common humanity.

CHAPTER III.

ETHICAL RELATIONS OF MAN TO BEAST.

Anthropocentric psychology and ethics. Teleological inferences from this postulate. Illustrations from Bernardin de Saint-Pierre and Gennadius. Its influence in checking the growth of science and the progress of hygiene. Natural phenomena regarded as portents. Astrology and horoscopy. Comets as warnings to mankind. Increase Mather's view. Bayle's ridicule of this theory. Notion that fruits and flowers exist only for man. The wasteful prodigality of Nature. Gray's sentiment on the subject. The real function of the colour and odour of plants. Schopenhauer on the anthropocentric principle in Judaism and Christianity. The Hebrew cosmogony. Man's dominion and its practical effects according to Shelley and Burns. Observation of Mrs. Jameson. Celsus's stricture indorsed by Dr. Thomas Arnold. Paley's defective definition of virtue. Bishop Butler on the immortality of animals. Opinion of Barclay. Philozoic philosophy of Henry Hallam. Denial of animals' rights by Catholic theologians and by Protestant writers on ethics. Influence of such theories upon modern legislation. Exposure by Samuel Plimsoll and Henry Bergh of the horrors of cattle transportation. "Horses cheaper than oats." American extravagance and recklessness. Erasmus Darwin's doctrine of the greatest possible happiness set aside by the science of evolution.

ETHNOCENTRIC geography, which caused each petty tribe to regard itself as the centre of the earth, and geocentric astronomy, which caused mankind to regard the earth as the centre of the universe, are conceptions that have been gradually outgrown and generally discarded—not, however, without leaving distinct and indelible traces of themselves in human speech and conduct. But this is not the case with anthropocentric psychology and ethics, which treat man as a being essentially different and inseparably set apart from all other sentient creatures, to which he is bound by no ties of mental affinity or moral obligation. Nevertheless, all these notions spring from the same root, having their origin in man's false and overweening conceit of himself as the member of a tribe, the inhabitant of a planet, or the lord of creation.

It was upon this sort of anthropocentric assumption that teleologists used to build their arguments in proof of the existence and goodness of God as shown by the evidences of beneficent design in the world. All their reasonings in support of this doctrine were based upon the theory that the final purpose of every created thing is the promotion of human happiness. Take away this anthropocentric postulate, and the whole logical structure tumbles into a heap of unfounded and irrelevant assertions leading to lame and impotent conclusions.

Thus Bernardin de Saint-Pierre states that garlic, being a specific for maladies caused by marshy exhalations, grows in swampy places, in order that the antidote may be easily accessible to man when he becomes infected with malarious disease. Also the fruits of spring and summer, he adds, are peculiarly juicy, because man needs them for his refreshment in hot weather; on the other hand, autumn fruits, like nuts, are oily, because oil generates heat and keeps men warm in winter. It is for man's sake, too, that in lands where it seldom or never rains there is always a heavy deposition of dew. If we can show that any product or phenomenon of Nature is useful to *us*, we think we have discovered its sufficient *raison d'être,* and extol the wisdom and kindness of the Creator; but if anything is harmful to us we can not imagine why it should exist. How much intellectual acuteness and learning have been expended to reconcile the fact that the moon is visible only a very small part of the time, with the theory that it was intended to illuminate the earth in the absence of the sun, for the benefit of its inhabitants!

Gennadius, a Greek presbyter, who flourished at Constantinople about the middle of the fifth century, remarks in his commentary on the first chapter of Genesis, that God created the beasts of the earth and the cattle after their kind on the same day on which he created man, in order that these creatures might be there ready to serve him.

But it would be superfluous to multiply examples of the influence of this anthropocentric idea as it has worked itself out in the history of mankind. Every science has had to encounter its opposition, and it has been a stumbling-block in the way of every effort to enlarge human knowledge and to promote human happiness. It has tended to check the progress of hygienic research and sanitary reform; for if man is of such exceptional importance that his conduct or misconduct can bring down epidemics upon whole communities and vast continents as visitations of divine wrath, whoever seeks to ward off or to stay these punishments is guilty of a sacrilegious attempt to parry the blow aimed at the wicked by the arm of the Almighty, and, by thus setting himself in antagonism to God, becomes in fact an ally and adversary of the devil. Thus vaccination was denounced, not on the ground taken by its present opponents, that it is useless as a preventive of smallpox and a prolific source of other diseases, but on account of its real or supposed prophylactic effectiveness, since it impiously wrenched from the hand of the Deity one of his most fatal weapons of retribution.

To what absurdities of presumption the anthropocentric conception has paved the way is evident from the belief, once universally entertained, that the sun, moon, and stars were placed in the firmament with express reference to man, and exerted a benign or baleful influence upon his destiny from the cradle to the grave. Owen Glendower's bombastic boast—

> . . . At my nativity
> The front of heaven was full of fiery shapes,
> Of burning cressets; and at my birth
> The frame and huge foundation of the earth
> Shaked like a coward—

was well answered by Hotspur: "Why, so it would have done at the same season if your mother's cat had but kittened, though yourself had ne'er been born." And yet this fulsome brag of the Welsh swashbuckler was only an extravagant statement of what the captious Henry Percy and his contemporaries all held to be virtually true. Poe embodies the same sentiment in his youthful poem, Al Aaraaf, and would fain preserve this brighter world of his fancy from the contagion of human evil—

> Lest the stars totter in the guilt of man.

Astrology and horoscopy, from which even the keen intellects of Kepler and Tycho de Brahe could not disentangle themselves, and to which the still more modern genius of Goethe paid a characteristic tribute in the story of his nativity, were only this anthropocentric conceit masquerading as science, and leaving vestiges of itself in such common words as "ill-starred" and "lunatic."

Comets were universally regarded as portents of disasters, sent expressly as warnings for the reproof and reformation of mankind; tempests and lightnings were feared as harbingers of divine wrath and instruments of punishment for human transgression. According to the Rev. Increase Mather, God took the trouble to eclipse the sun in August, 1672, merely to prognosticate the death of the President of Harvard College and of two colonial governors, all of whom "died within a twelvemonth after." This is but a single example of the wide prevalence and general acceptance of a popular superstition constantly tested and easily proved by the logical fallacy *post hoc ergo propter hoc*. Bayle, in his Divers Thoughts on Comets (Pensées Diverses sur les Comètes), ridicules the foolish pride and vanity of man, who imagines that "he can not die without disturbing the whole course of Nature and compelling the heavens to put themselves to fresh expense in order to light his funeral pomp."

Not only were the fruits of the earth made to grow for human sustenance, but the flowers of the field were supposed to bud and blossom, putting on their gayest attire and emitting their sweetest perfume, solely as a contribution to human happiness; and it was deemed one of the mysteries and mistakes of Nature, never too much to be puzzled over and wondered at, that these things should spring up and expend their beauty and fragrance in remote places untrodden by the foot of man. Gray expresses this feeling in the oft-quoted lines:

> Full many a flower is born to blush unseen,
> And waste its sweetness on the desert air.

Science has finally and effectually taken this conceit out of man by showing that the flower blooms not for the purpose of giving him agreeable sensations, but for its own sake, and that it presumed to put forth sweet and beautiful blossoms long before he appeared on the earth as a rude cave-haunting and flint-chipping savage.

The colour and odour of the plant are designed not so much to please man as to attract insects, which promote the process of fertilization and thus insure the preservation of the species. The gratification of man's æsthetic sense and taste for the beautiful does not enter into Nature's intentions; and although the flower may bloom unseen by any human eye, it does not on that account waste its sweetness, but fully accomplishes its mission, provided there is a bee or a bug abroad to be drawn to it. That the fragrance and variegated petals are alluring to a vagrant insect is a condition of far more importance in determining the fate of the plant than that they should be charming to man.

Plants, on the other hand, which depend upon the force of the wind for fructification, are not distinguished for beauty of colour or sweetness of odour, since these qualities, however agreeable to man, would be wasted on the wind. This is an illustration of the prudent economy of Nature, who never indulges in superfluities or overburdens her products with useless attributes; but the test of utility which " great creating Nature" sets up in such cases is little flattering to man, and has no reference to his tastes and susceptibilities, but is determined solely by the serviceableness of certain qualities to the plant itself in the struggle for existence.

According to Schopenhauer, anthropocentric egoism is a fundamental and fatal defect in the psychological and ethical teachings of both Judaism and Christianity, and has been the source of untold misery to myriads of sentient and highly sensitive organisms. "These religions," he says, "have unnaturally severed man from the animal world, to which he essentially belongs, and placed him on a pinnacle apart, treating all lower creatures as mere things; whereas Brahmanism and Buddhism insist not only upon his kinship with all forms of animal life, but also upon his vital connection with all animated Nature, binding him up into intimate relationship with them by metempsychosis."

In the Hebrew cosmogony there is no continuity in the process of creation, whereby the genesis of man is in any wise connected with the genesis of the lower animals. After the Lord God, by his fiat, had produced beasts, birds, fishes, and creeping things, he ignored all this mass of protoplastic and organic material, and took an entirely new departure in the production of man, whom he formed out of the dust of the ground. Science shows him to have been originally a little higher than the ape, out of which he was gradually and painfully evolved; Scripture takes him out of his environment, severs him from his antecedents, and makes him a little lower than the angels. Upon the being thus arbitrarily created absolute dominion is conferred over every beast of the earth and every fowl of the air, which are to be to him "for meat." They are given over to his supreme and irresponsible control, without the slightest injunction of kindness or the faintest suggestion of any duties or obligations toward them.

Again, when the earth is to be renewed and replenished after the deluge, the same principles are reiterated and the same line of demarcation is drawn and even deepened. God blesses Noah and his sons, bids them "be fruitful and multiply," and then adds, as regards the lower animals: "The fear of you and the dread of you shall be upon every beast of the earth and upon every fowl of the air, upon all that moveth upon the earth, and upon all the fishes of the sea; into your hand are they delivered. Every moving thing that liveth shall be meat for you; even as the green herb have I given you all things."

This tyrannical mandate is not mitigated by any intimation of the merciful manner in which the human autocrat should treat the creatures thus subjected to his capricious will. On the contrary, the only thing that he is positively commanded to do with reference to them is to eat them. They are to be regarded by him simply as food, having no more rights and deserving no more consideration as means of sating his appetite than a grain of corn or a blade of grass.

The practical working of this decree has been summed up by Shelley, with his wonted force and succinctness, when he says, "The supremacy of man is, like Satan's, a supremacy of pain." Burns regrets the fatal effect of the sovereignty thus conferred upon the human race in destroying the mutual sympathy and confidence which should exist between the lord of creation and the lower animals in the lines addressed To a Mouse, on turning her up in her Nest with the Plough, November, 1785:

> I'm truly sorry man's dominion
> Has broken Nature's social union,
> An' justifies that ill opinion
> Which makes thee startle
> At me, thy poor earth-born companion,
> An' fellow-mortal.

In the subsequent annals of the world we have ample commentaries on this primitive code written in the blood of helpless, innocent, and confiding creatures, which, although called dumb and incapable of recording their sufferings, yet

> . . . have long tradition and swift speech,
> Can tell with touches and sharp-darting cries
> Whole histories of timid races taught
> To breathe in terror by red-handed man.

Indeed, ever since Abel's firstlings of the flock were more acceptable than Cain's bloodless offerings of the fruits of the fields, priests have performed the functions of butchers, converting sacred shrines into shambles in their endeavours to pander to the gross appetites of cruel and carnivorous gods. Cain's offering was rejected, says Dr. Kitto, because "he declined to enter into the sacrificial institution." In other words, he would not shed the blood of beasts to gratify the Lord—a refusal which we can not but regard as exceedingly commendable in Adam's first-born.

"I do not remember," observed Mrs. Jameson, "ever to have heard the kind and just treatment of animals enforced on Christian principles or made the subject of a sermon." George Herbert was a man of gentle spirit and ready hand for the relief of all forms of human distress, and in his book entitled A Priest to the Temple, or the Country Parson, lays down rules and precepts for the guidance of the clergyman in all relations of life, even to the minutest circumstances and remotest contingencies incident to parochial care. But this tender-hearted man does not deem it necessary for the parson to take the slightest interest in animals, and does not utter a word of counsel as to the manner in which his parishioners should be taught their duties toward the creatures so wholly dependent upon them. Indeed, no treatise on pastoral theology ever touches this topic, nor is it ever made the theme of a discourse from the pulpit, or of systematic instruction in the Sunday school.

Neither the synagogue nor the church, neither sandedrin nor ecclesiastical council, has ever regarded this subject as falling within its scope, and sought to inculcate as a dogma or to enforce by decree a proper consideration for the rights of the lower animals. One of the chief objections urged by Celsus more than seventeen centuries ago against Christianity was that it "considers everything as having been created solely for man." This stricture is indorsed by Dr. Thomas Arnold, of Rugby, who also animadverts on the evils growing out of the anthropocentric character of Christianity as a scheme of redemption and a system of theodicy. "It would seem," he says, "as if the primitive Christian, by laying so much stress upon a future life in contradistinction to this life, and placing the lower creatures out of the pale of hope, placed them at the same time out of the pale of sympathy, and thus laid the foundation for this utter disregard of animals in the light of our fellow-creatures. The definition of virtue among the early Christians was the same as Paley's—that it was good performed for the sake of insuring eternal happiness—which of course excluded all the so-called brute creatures. Kind, loving, submissive, conscientious, much-enduring, we know them to be; but because we deprive them of all stake in the future, because they have no selfish, calculated aim, these are not virtues; yet if we say 'a vicious horse,' why not say 'a virtuous horse'?"

We are ready enough, adds Dr. Arnold, to endow animals with our bad moral qualities, but grudge them the possession of our good ones. The Germans, whose natural and hereditary sympathy with the brute creation is stronger than that of any other Western people, speak of horses as "*fromm*," pious, not in the religious, but in the primary and proper sense of the word, meaning thereby kind and docile. The English "gentle" and the French "*gentil*," which are used in the same connection, refer to good conduct as the result of fine breeding.

Archdeacon Paley's definition of virtue, to which Dr. Arnold adverts, is essentially anthropocentric and intensely egoistic. "Virtue," he says, "is the doing good to mankind in obedience to the will of God, for the sake of everlasting happiness." In order to be virtuous, according to this extremely narrow and wholly inadequate conception of virtue, we must, in the first place, do good to mankind, our conduct toward the brute creation not being taken into the account; secondly, our action must be in obedience to the will of God, thus ruling out all generous impulses originating in the spontaneous desire to do good; thirdly, we must have an eye single to our own supreme personal advantage—in other words, our conduct must be utterly selfish, spring not merely from momentary pleasure or temporary profit, but from far-seeing calculations of the effect it may have in securing our eternal happiness. Thus the virtuous man becomes the incarnation of the intensest self-love and self-seeking, and virtue the synonym of excessive venality. From a moral point of view, there is no greater merit in "otherworldliness" than in worldliness, and no reason why the endeavour to attain personal happiness in a future life should differ in quality from the effort to make everything minister to our personal happiness in the present life.

"The whole subject of the brute creation," says Dr. Arnold, "is to me one of such painful mystery that I dare not approach it." The mental distress experienced in such cases arises from the fact that the subject is approached from the wrong side and surveyed from a false point of view. Traditional theology and anthropocentric ethics are brought into conflict with the better impulses of a broad and generous nature and the sharp antagonism could hardly fail to be a source of perplexity and pain. "Charity," says Lord Bacon, "will hardly water the ground, where it must first fill a pool"; and of all pools the hardest to fill is that which is dug in the dry, gravelly soil of human egotism.

Theocritus, the father of Greek idyllic poetry, represents Hercules as exclaiming, after he had slain the Nemean lion, "Hades received a monster soul"; and he saw nothing incongruous in the spirit of the dead beast joining the company of the departed spirits of men in the lower world. Sydney Smith says, in speaking of the soul of the brute, "To this soul some have impiously allowed immortality." Why such a belief should be deemed impious it is difficult to discover. The question which the psychologist has to consider is not whether the doctrine is impious, but whether it is true. No scientific opinion has ever been advanced that has not seemed impious to some minds, and been denounced and persecuted as such by ecclesiastical authorities.

Bishop Butler, on the contrary, in his work on The Analogy of Religion, Natural and Revealed, to the Constitution and Course of Nature, declares that "we can not find anything throughout the whole analogy of Nature to afford us even the slightest presumption that animals ever lose their living powers." He admits that his argument in support of the doctrine of a future life proves the immortality of brutes as well as that of man, and thus recognises their spiritual kinship.

An eminent Scotch physician and anatomist, Dr. John Barclay, in his Inquiry into the Opinions, Ancient and Modern, concerning Life and Organization (1825), urges the probable immortality of the lower animals, which, he thinks, are "reserved, as forming many of the accustomed links in the chain of being, and by preserving the chain entire, contribute in the future state, as they do here, to the general beauty and variety of the universe, a source not only of sublime but of perpetual delight." The author seems to infer the continued existence of the brute creation from the fact that it forms an essential part of universal being, and that its total disappearance would mar the perfection of the next world, which should be more perfect than this world. He assumes, however, that the lower animals are endowed with immortality, not so much from psychological necessity or for their own sake as sentient and intelligent creatures, as for man's sake, in order that their presence may minister to his pleasure by forming an attractive feature in the heavenly landscape. It is, therefore, solely from anthropocentric considerations that they are granted this lease of eternal life; just as "the poor Indian" is represented by the poet as looking forward to the possession of happy hunting fields after death, where he may follow with keener enjoyment his favourite pursuit, and "his faithful dog shall bear him company."

More than fifty years ago Henry Hallam made the following observations, which are remarkable as an anticipation of the ethical corollary to the doctrine of evolution: "Few at present, who believe in the immortality of the human soul, would deny the same to the elephant; but it must be owned that the discoveries of zoölogy have pushed this to consequences which some might not readily adopt. The spiritual being of a sponge revolts a little our prejudices; yet there is no resting place, and we must admit this or be content to sink ourselves into a mass of medullary fibre. Brutes have been as slowly emancipated in philosophy as some classes of mankind have been in civil polity; their souls, we see, were almost universally disputed to them at the end of the seventeenth century, even by those who did not absolutely bring them down to machinery. Even within the recollection of many, it was common to deny them any kind of reasoning faculty, and to solve their most sagacious actions by the vague word instinct. We have come in late years to think better of our humble companions; and, as usual in similar cases, the preponderant bias seems rather too much of a levelling character." During the half century that has elapsed since these words were written, not only has zoölogy made still greater progress in the direction indicated, but a new science of zoöpsychology has sprung up, in which the mental traits and moral qualities of the lower animals have been, not merely recorded as curious and comical anecdotes, but systematically investigated and philosophically explained. In consequence of this radical change of view, human society in general has become more philozoic, not upon religious or sentimental but upon strictly scientific grounds, and developed a sympathy and solidarity with the animal world, having its sources less in the tender and transitory emotions of the heart than in the profound and permanent convictions of the mind.

In an essay published a few years ago in The Dublin Review (October, 1887, p. 418), the Right Rev. John Cuthbert Hedley, Bishop of Newport and Menevia, asserts that animals have no rights, because they are not rational creatures and do not exist for their own sake. "The brute creation have only one purpose, and that is to minister to man, or to man's temporary abode." This is the doctrine set forth more than six centuries ago by Thomas Aquinas, and recently expounded by Dr. Leopold Schutz, professor in the theological seminary at Trier, in an elaborate work entitled The Socalled Understanding of Animals or Animal Instinct. This writer treats the theory of the irrationality of brutes as a dogma of the Church, denouncing all who hold that the mental difference between man and beast is one of degree, and not of kind, as "enemies of the

Christian faith"; whereas those who cling to the old notion of instinctive or automatic action in explaining the phenomena of animal intelligence are extolled as "champions of pure truth."

In an article on The Lower Animals, in the Catholic Dictionary of W. E. Addis and T. Arnold, published in 1884, it is maintained that "the brutes are made for man, who has the same right over them which he has over plants and stones," and that it is lawful for him to put them to death and to torment them "even for the purposes of recreation." A similar view is taken by Philip Austin in a volume on Our Duty to Animals (London, 1885), in which the author, treating the subject "in the light of Christian philosophy," comes to the conclusion "that kindness to the brutes is a mere work of supererogation."

If it was the Creator's intention that the lower animals should minister to man, the divine plan has proved to be a failure, since the number of animals which, after centuries of effort, he has succeeded in bringing more or less under his dominion is extremely small. Millions of living creatures fly in the air, crawl on the earth, dwell in the waters, and roam the fields and the forests, over which he has no control whatever. Not one in twenty thousand is fit for food, and of those which are edible he does not actually eat more than one in ten thousand. In explanation of this lack of effectiveness in the enforcement of a divine decree, it has been asserted that man lost his dominion over the lower world to a great extent when he lost dominion over himself; but this view is wholly untenable even from a biblical standpoint, inasmuch as the promise of universal sovereignty was renewed after the deluge and expressed in even stronger terms than before the fall.

Dugald Stewart admits "a certain latitude of action, which enables the brutes to accommodate themselves in some measure to their accidental situations." In this arrangement he sees a design or purpose of "rendering them, in consequence of this power of accommodation, incomparably more serviceable to our race than they would have been if altogether subjected, like mere matter, to the influence of regular and assignable causes." Of the value of this power of adaptation to the animal itself in the struggle for existence the Scotch philosopher had no conception.

In the great majority of treatises on moral science, especially in such as base their teachings on distinctively Christian tenets, there is seldom any allusion to man's duty toward animals. Dr. Wayland, who has perhaps the most to say on this point, sums up his remarks in a note apologetically appended to the body of his work. He denies them the possession of "any moral faculty," and declares that in all cases "our right is paramount and must extinguish theirs." We are to treat them kindly, feed and shelter them adequately, and "kill them with the least possible pain." To inflict suffering upon them for our amusement is wrong, since it tends to harden men and render them brutal and ferocious in temper.

Dr. Hickok takes a similar view and broadly asserts that "neither animate nor inanimate Nature has any rights," and that man is not bound to it "by any duties for its own sake. . . . In the light of his own worthiness as end, . . . he is not permitted to mar the face of Nature, nor wantonly and uselessly to injure any of her products." Maliciously breaking a crystal, defacing a gem, girdling a tree, crushing a flower, painting flaming advertisements on rocks, and worrying and torturing animals are thus placed in the same category as acts tending to degrade man ethically and æsthetically, rendering him coarse and rude, and making him not only a very disagreeable associate, but also, in the long run, "an unsafe member of civil society." These things are considered right or wrong solely from the standpoint of their influence upon human elevation or degradation. "Nature possesses no product too sacred for man. All Nature is for man, not man for it."

The same opinion is held by the Jesuit, Victor Cathrein, who, in a recently published review of Bregenzer's Thier-Ethik (Stimmen aus Marien-Lach, February 7, 1895, p. 164), denies that man has any duties toward the lower animals, and asserts that any cruelty he may inflict upon them involves no moral wrong different in kind from that which he commits in wantonly tearing or dirtying his own clothes. According to this doctrine, animals have no more rights than inanimate objects, and it is no worse from an ethical point of view to flay the forearm of an ape or lacerate the leg of a dog than to rip open the sleeve of a coat or rend a pair of pantaloons. The plain statement of such a theory is its sufficient refutation, and we doubt whether even such a severe dogmatist and uncompromising champion of Catholic principles as Victor Cathrein, S. J., would be able to witness all these operations with equal equanimity.

Man is as truly a part and product of Nature as any other animal, and this attempt to set him up on an isolated point outside of it is philosophically false and morally pernicious. It makes fundamental to ethics a principle which once prevailed universally in politics and still survives in the legal fiction that the king can do no wrong. Louis XIV of France firmly believed himself to be the rightful and absolute owner of the lives and property of his subjects. He held that his rights as monarch were paramount and extinguished theirs, that they possessed nothing too sacred for him, and the leading moralists and statists of his day confirmed him in this extravagant opinion of his royal prerogatives. All the outrages which the mad Czar, Ivan the Terrible, perpetrated on the inhabitants of Novgorod and Moscow, man has felt and for the most part still feels himself justified in inflicting on domestic animals and beasts of venery.

It is only within the last century that legislators have begun to recognise the claims of brutes to just treatment and to enact laws for their protection. Torturing a beast, if punished at all, was treated solely as an offence against property, like breaking a window, barking a tree, or committing any other act known in Scotch law as "malicious mischief." It was regarded, not as a wrong done to the suffering animal, but as an injury done to its owner, which could be made good by the payment of money. Not until a little more than a hundred years ago was such an act changed from a civil into a criminal offence, for which a simple fine was not deemed a sufficient reparation. It was thus placed in the category of crimes which, like arson, burglary, and murder, are wrongs against society, for which no pecuniary restitution or compensation can make adequate atonement.

Even this legislative reform is by no means universal.

The criminal code of the German Empire still punishes with a fine of not more than fifty thalers any person "who publicly, or in such wise as to excite scandal, maliciously tortures or barbarously maltreats animals." This sort of cruelty is classified with drawing plans of fortresses, using official stamps and seals, and putting royal or princely coats of arms on signs without permission, making noises which disturb the public peace, and playing games of hazard on the streets or market places. The man is punished, not because he puts the animal to pain, but because his conduct is offensive to his fellow-men and wounds their sensibilities. The law sets no limit to his cruelty, provided he may practise it in private.

Again, in all enactments regulating the transportation of live stock our legislation is still exceedingly defective. The great majority of people have no conception of the unnecessary and almost incredible suffering inflicted by man upon the lower animals in merely conveying them from one place to another in order to meet the demands of the market. It is well known that German shippers of sheep to England often lose one third of their consignment by suffocation, owing to overcrowding and imperfect ventilation. Beasts are still made to endure all the horrors to which slavers were once wont to subject their cargoes of human chattels in stifling holds on the notorious "middle passage." (Some conception of the cruelties involved in this traffic may be obtained by reading Samuel Plimsoll's little volume entitled Cattle Ships, published in London in 1890.)

The late Henry Bergh states that the loss on cattle by "shrinkage" in transporting them from the Western to the Eastern portion of the United States is from ten to fifteen per cent. The average shrinkage of an ox is one hundred and twenty pounds, and that of a sheep or hog from fifteen to twenty pounds; and the annual loss in money arising from this cause is estimated at more than forty million dollars. The amount of animal suffering which these statistics imply is fearful to contemplate. Here and there a solitary voice is heard in our legislative halls protesting against the horrors of this traffic, but so powerful is the lobby influence of wealthy corporations that no law can be passed to prevent them. Not a word ever falls from the pulpit in rebuke of such barbarity; meanwhile the railroad magnates pay liberal pew rents out of the profits, and listen with complacency one day in the week to denunciations of Jeroboam's idolatry and the wicked deeds of Ahab and Ahaziah, as recorded in the chronicles of the kings of Israel.

The horse, one of the noblest and most sensitive of domestic animals, is put to all kinds of torture by docking, pricking, clipping, peppering, and the use of bearing reins solely to gratify human vanity. As a reward for severe and faithful toil he is often fed with unwholesome and insufficient fodder on the economical principle announced by the manager of a New York tramway that "horses are cheaper than oats." It is an actual fact, verified by Henry Bergh, that the horses of this large corporation were fed on a mixture of meal, gypsum, and marble dust, until the Society for the Prevention of Cruelty to Animals interfered and finally succeeded in putting a stop to the practice.

The Americans, as a people, are notorious for the recklessness with which they squander the products of

Nature, of which their country is so exceedingly prolific. This extravagance extends to all departments of public, social, and domestic life. No land less rich in material resources could have borne for any length of time the wretched mismanagement of its finances to which the United States has been subjected ever since and even before the close of the civil war. There is not a government in Europe that would not have been broken down and rendered bankrupt by the tremendous and wholly unnecessary strain put upon it by crass ignorance of the most elementary principles of finance and demagogical tampering with the public credit. The same wasteful spirit involves also, as we have seen, immense suffering to animals on the part of soulless and unscrupulous corporations, in which intense greed of gain is not mitigated by the influence of individual kindness, and by which horses are treated as mere machines, to be worked to their utmost capacity at the smallest expense, and neat cattle as so much butcher's meat to be brought to market in the quickest and cheapest manner.

Erasmus Darwin, in his Phytologia, or the Philosophy of Agriculture and Gardening (London, 1800), endeavours to vindicate the goodness of God in permitting the destruction of the lower by the higher animals on the ground that "more pleasurable sensation exists in the world, as the organic matter is taken from a state of less irritability and less sensibility and converted into a greater." By this arrangement, he thinks, the supreme sum of possible happiness is secured to sentient beings. Thus it may be disagreeable for the mouse to be caught and converted into the flesh of the cat, for the lamb to be devoured by the wolf, for the toad to be swallowed by the serpent, and for sheep, swine, and kine to be served up as roasts and ragouts for man; but in all such cases, he argues, the pain inflicted is far less than the amount of pleasure ultimately procured. But how is it when a finely organized human being, with infinite capabilities of happiness in its highest forms, is suddenly transmuted into the bodily substance of a boa constrictor or a tiger? No one will seriously assert that the drosera, *Dionæa muscipula,* and other insectivorous and carnivorous plants are organisms superior in sensitiveness to those which they devour, or that this transformation of animal into vegetable structure increases the sum of pleasurable sensation in the world. The doctrine of evolution, which regards these antagonisms as mere episodes in the universal struggle for existence, has forever set aside this sort of theodicy and put an end to all teleological attempts to infer from the nature and operations of creation the moral character of the Creator.

CHAPTER IV.

METEMPSYCHOSIS.

Universality of the belief in the transmigration of souls. Conception of immortality among primitive tribes. Strong faith of the savage. Filial affection as exemplified by parricide. Persistence of the dogma of metempsychosis. Traces of it in Judaism and Christianity. Elect Israelitic souls. Metempsychosis taught by the Manichæans and used by Origen to explain divine predestination. Pre-existence held by Pythagoras, Plato, and other Greek philosophers and assumed to be true by Jesus. Augustine's commentary on the Golden Ass of Appuleius. Goethe's confession. Appuleius and Czeslav Czynski as hypnotizers. Relation of zoölatry to metempsychosis. Animals as incarnations of ghosts and demons. Metempsychosis as the metaphorical expression of human aspiration and evolution. The spiritual law of like seeking like in the predetermination of character. Indian conception of fate and free will illustrated by modern statistics of crime, suicide, and other social phenomena. Plato's theory of the origin of intuitive knowledge. "Essential spissitude." Lessing on the possibility of more than five senses in man. Neo-Lamarckism. Pervading influence of pantheism in the Orient. Indian atheists. Brahmanical and Buddhistic eschatology: absorption and extinction of the individual soul as the radical cure of egotism. Paul and Nanak. The pantheistic compared with the Christian scheme of salvation from an ethical point of view. Transmigration of souls and transmutation of species. Conservation of force and imperishableness of spirit. Thomas Aquinas's untenable distinction between human and sub-human souls. Moral bearing of metempsychosis. Orientals in their treatment of animals not always true to their religious precepts. Protection of animals as property. Panpsychic philosophy. Societies for the prevention of cruelty to animals. Hospitals for beasts in India. Monier Williams's description of the Panjara Pol. Mantegazza's account of such an institution. The "Bai Sakarbai." King Thibo and Barnum. European hospitals for animals. The New York Veterinary Hospital. Lecky's observations. Oriental and Occidental treatment of animals contrasted. Lack of pertinent biblical texts. Quandary of a Protestant parson. Deficiencies of Hebrew and Christian Scriptures. Animals in hagiology. Ecclesiastical excommunication of animals. Festivals of St. Anthony in Rome and of St. Leonard in Tölz. Legends of St. Francis of Assisi. Indifference of the Catholic Church to the sufferings of animals. Dictum of Pius the Ninth. Its practical application by Italians. Spanish bullfights and the popes. Beneficent influence of evolutionary science and comparative psychology upon the humane treatment of animals.

IT is especially in man's conception of his relations to the lower animals and of the character and degree of their psychical development and mental endowment that anthropocentric prejudices and prepossessions continue to exert a perverting and pernicious influence.

Opposed to this tendency, both as a philosophical principle and in its bearings on practical ethics, is the doctrine of the transmigration of souls. If the truth of a tenet may be determined by the majority of suffrages in its favour, if the validity of a theory bears any proportion to the number of persons who have accepted it and found comfort and consolation in it, if the famous test *quod semper, quod ubique, quod ab omnibus*, which the Romish Church has made the criterion of its own claim to catholicity, has any force or fitness as furnishing a ground of belief, it would be difficult to discover among the multiform creeds of mankind any doctrine resting upon a broader and firmer foundation than that which is known as metempsychosis.

Indeed, if such indorsement is to be regarded as any proof of genuineness, this theory may be said, without exaggeration, to be stamped with the seal of almost universal consent, since it has been found to be inherent in or engrafted upon nearly every known school of philosophy and system of religion, and to have been held, in some of its varied forms, by men in all ages, in all lands, in all conditions of life, and in all stages of barbarism and civilization.

The belief in the transmigration of souls and in their progressive improvement through successive stages of incarnation is common to the aboriginal tribes of every land, and may be regarded as the earliest and most general form in which the conception of immortality takes expression. To the mind of the primitive man the idea of the continued existence of the soul in a disembodied state is utterly incomprehensible, and would be equivalent to its permanent extinction. After having come in contact with Europeans and learned to appreciate their superiority, the negro's ideal of immortality is to animate, after death, the body of a white man. One of the strongest incentives of the savage to distinguish himself in battle is the hope of being rewarded for his prowess by being born again into a higher tribal position as a mighty chieftain or a powerful medicine man. So firm is his conviction of this possibility, that he often courts danger and craves death in order to better his condition by a new birth.

Culture is critical and sceptical in its relations to the unseen world and touching all that lies beyond the bourn of the present life. Only barbarism is capable of begetting the intense and implicit faith that never questions the words of the priest or suspects the wiles of the wizard. This crass credulity is characteristic of infant intelligence, and disappears with the mental growth and maturity of the race. Where it exists in full force it always produces a fearlessness bordering on fanaticism, as in the soldiery of the Sikh Guru and the Mohammedan Mahdi, or as in the case of the Congo negress, who put such perfect confidence in the protecting power of her fetiches, that she unhesitatingly placed her foot on a block and permitted it to be struck off with an ax, and could hardly believe that amulets and charms had failed to prevent the natural effects of the blow. The same amount of superstitious assurance in a civilized man would be regarded as conclusive proof of his insanity. We have an example of this kind in the sect known as the "peculiar people," who, not having outgrown the healing methods enjoined and employed by the Christian Church in the first century, are constantly coming into conflict with the hygienic regulations established and enforced by Christian governments in the nineteenth century.

With what unwavering trust the old German warrior had his weapons, his wives, his horses, and his slaves buried in his dolmen, never doubting that they would go with him and be ready for his service in the next world! He was the best and foremost man of his time; but should one of his descendants of to-day attempt to express in like manner his firm faith in the immortality of the soul, he would be denounced as a dangerous religious "crank," and summarily arrested by the police. In Polynesia it was thought to be the duty of the child to put the parents to death as soon as the physical powers began to show symptoms of decay. The purpose of the parricide was not to rid himself of a burden, but sprung solely from feelings of filial affection and prescriptive obligation, and from the desire that his parents might escape the infirmities of old age and enter in full vigour upon the future life. The parents consented to the act and were happy in the prospect of speedy rejuvenation in the realms of the blest. So, too, among the Battas of Sumatra, a gentle and kindly race, the father, when he feels the signs of approaching old age, begs his sons to kill and eat him. On the day appointed for the performance of this filial duty, the old man climbs up into a tree, round which his sons stand, beating upon the trunk and singing a sort of dirge, the burden of which is: "The season has come, the fruit is ripe and must fall." Thereupon the old man descends, and is solemnly slain and lovingly devoured. But where is the Christian, however zealous and sincere, who would run so great a risk, or whose faith in the resurrection of the dead and the life everlasting would stand such a terrible test? If he could be found, his proper place would be, not in the sanctuary of the saints, but in an asylum for the insane.

Metempsychosis is not merely a dogma of the past or lingering survival of primitive beliefs. It is still a living psychological principle and practical precept of religion, and numbers its adherents by millions, including all grades of enlightenment, from the African or Australian savage to the Oriental sage, and all degrees and developments of spiritual aspiration, from the rudest rubbish worship of the Loango fetichist to the most refined mysticism of the European philosopher. It appears with the earliest dawn of Indian speculation, and pervades the whole vast, subtile, and complicated web of Brahmanical metaphysics. It is the central and sustaining root of that widespreading banyan of Buddhistic ethics, which extends its ample and hospitable shade over the entire realm of animated nature, and gives impartial shelter and protection to every form of animal life. It constituted an integral part of the priestly wisdom of Egypt, fragments of which have been preserved and transmitted to us in the so-called Book of the Dead. The custom of embalming the deceased grew out of the belief that the souls of the departed would, after long wanderings and numerous transformations, return to re-inhabit their human bodies, and undergo again in this form various trials and purifications preparatory to a final and eternal union with Osiris. According to Herodotus (ii, 123), this transmigration embraced in its circuit the principal animals of the earth, the sea, and the air, and took three thousand years for its accomplishment. Plastic and pictorial illustrations of this doctrine are found on Egyptian monuments and papyri, as, for example, where the soul of a glutton is represented as being borne to Hades in the form of a hog.

Even the Jews, notwithstanding the essential inconsistency of the theory of transmigration with their cosmogony and the prevailing spirit of their sacred scriptures, borrowed it, together with the conception of a future life, from their conquerors during the Babylonian captivity; and a tenderer feeling toward the lower animals is clearly perceptible among them in consequence of their long and intimate contact with Assyrian and Persian ideas and habits of thought. It finds, therefore, as one would naturally expect, its most frequent expression and fullest unfolding in the apochryphal and exegetical literature of the Hebrews, while in the so-called canonical writings there are only faint and comparatively few traces of it. Thus the author of the Book of Wisdom says of himself: "I was a well-conditioned child and had received a good soul; and, since I was still good, I went into an immaculate body." The Cabala declares still more emphatically that "all souls are subject to the trials of transmigration." The Talmud reiterates the same thought. Many of the most eminent rabbis taught that the souls of men are sometimes condemned to inhabit the bodies of women as a punishment for sins of effeminacy and for mean and unmanly deeds, thus producing such monstrosities as amazons and viragoes. They also ascribed barrenness in women to the penal possession of a male soul, in which case she was enjoined to entreat the Lord to pardon her offence, committed in a former body, and graciously to grant her the power of childbearing by endowing her with scintillations of a female soul. They maintained, furthermore, that in the beginning God created a certain number of Jew souls as his elect, and that these souls constantly return to animate the bodies of successive generations of the chosen people, and will remain a select source of spiritual supply as long as the seed of Abraham continues to dwell upon the earth. This is also supposed to account for the rare persistency of race peculiarities which characterizes Israelites. According to this theory, Jew souls never stray into Gentile bodies, though they are frequently made to atone for their sins by becoming incarnate in beasts. It is also stated that when this process of transmigration and purification is complete, and every Jew soul animates the body of a just Jew, then the end of the world will come. It might seem to many that to make this final event dependent upon such a remarkable concurrence of circumstances and happy condition of things would be equivalent to its indefinite postponement.

Of all Christian sectaries, the Manichæans were most considerate and careful of the lower animals, and this kindly attitude of mind was due, in a large measure, to the strong admixture of Oriental ideas in their system of belief. They held that the souls of men undergo transformations, passing successively into the bodies of beasts and birds and reptiles, partly as a method of punishment and partly as a means of growth and a process of purgation from the spiritually contaminating lusts of the flesh. Finally, after having been bathed in the sacred water of the moon and burned in the sacred fire of the sun and thus cleansed from all traces of material pollution, they become fit for admission, as pure spiritual essence, into the world of everlasting light.

Metempsychosis was also taught by Origen, who found in this doctrine a convenient master-key to the hidden meaning of many strange events and difficult passages of Scripture. Thus he explains the prenatal struggle of Esau and Jacob in the womb of Rebekah as the revival and continuation of a pre-existent enmity between them. The Lord likewise ordained Jeremiah to be a prophet unto the nations, while he was yet unborn, because, as is expressly stated, he had known and tried him in a previous state of being. Predestination, in Origen's opinion, could be brought into harmony with divine justice only on the same principle. God's discrimination between persons before their birth, foreordaining the one to everlasting life and the other to everlasting death, he held to be an outrageous wrong and an act of unpardonable favouritism, unless justified by their known character and antecedent conduct and their good or evil propensities as manifested in a former existence. To this most genial and thoughtful theologian of the Eastern church the rigorous dogma of the divine decrees, as implied in Paul's metaphor of the potter and the clay, was tyrannous and atrocious, and he took refuge from it in the intricate mazes of Buddhistic psychology and ethics.

This view of predestination would relieve it, in a certain degree, of its arbitrary and unjust character and establish a causal connection between the past conduct of the person, rewarded or punished, and his future condition. The divine decree would resolve itself into fate, and a man's fate, says an Indian sage, is the resultant of his deeds committed in a former body.

Origen held, too, that the story of the garden of Eden is an account of the life of our first parents in a previous state of existence, in which they fell into sin through disobedience and were condemned to dwell in human bodies. The passage in which God is said to have made "coats of skins" for Adam and Eve "and clothed them," means that he vestured them with mortal flesh as a punishment for their transgression. According to this theory, which is as radically pessimistic as any tenet of Buddhism, man became incarnated and, as it were, incarcerated in his present physical form in consequence of a curse, and his whole life on earth is that of a convict in a penal colony, and the chief end of his endeavours and aspirations should be to obtain pardon and redemption by winning the favour of the Almighty King who placed him in this durance vile.

Pythagoras claimed to have a distinct recollection of his pre-existent actions and experiences. Socrates maintained that all acquisition of knowledge or learning is nothing but remembering—ἡ μάθησις οὐκ ἄλλο τι ἢ ἀνάμνησις. Plato would condemn all cowardly and effeminate men, such as dandies and dudes, to be re-born as women; frivolous and flighty and feather-brained persons, to become birds; those who neglect the study of philosophy and seek only sensual indulgence, to be transformed into beasts; and the dull and foolish, to descend to the lower level of fishes and mollusks. He states, however, that Orpheus reappeared as a swan, and Thamyris as a nightingale at their own request a thousand years after their death. Aristotle held that the souls of poets are fond of taking bodily form again as cygnets; and Horace celebrates in an ode (ii, 20) his own apokyknosis or swan re-incarnation, and rejoices in the prospect of putting on feathers and soaring through the argent fields of air on the twofold pinions of bird and bard.

Jesus and his disciples seem to have assumed, at least on one occasion, that a person might suffer afflictions in this body as a punishment for sins committed by him before his birth or in a former state of conscious and responsible existence. Thus, we are told that as they passed by and saw a man that was blind from his birth, "his disciples asked him, saying, Master, who did sin, this man or his parents, that he was born blind?" (John ix, 2). In his reply Jesus admits implicitly that both of these hypotheses are legitimate and adequate to account for the given phenomenon, and only denies their applicability to this particular case of physical infirmity. In the same connection, Jesus asserts concerning himself that he existed before Abraham was. The statement in this passage and in others of a similar character, which have been usually interpreted as referring to his eternal Godhead, could be far more easily and rationally explained as expressions of his belief in the pre-existence of the soul.

St. Augustine maintained that men might be changed into beasts by sorcery, and even suggested, not sarcastically, but seriously, that the Golden Ass of Apuleius might be autobiographical, the author describing his adventures in that state, whereinto, by the evil arts of an enchantress, he had been actually "translated," like Bottom in the play. "In certain districts of Italy," he adds, "such occurrences are quite frequent. The women, who tend the herds, prepare with magic rites a kind of cheese, which they give to travellers to eat and thus change them into beasts of burden, in which shape they are made to bear heavy loads and perform other onerous tasks."

According to one tradition, Apuleius wished to become a creature with wings, but the malicious witches Meroe and Panthia, who seem to have been fond of playing practical jokes at others' expense, rubbed him with an ointment which changed him into an ass, doubtless thinking that this would be the easiest and most natural metamorphosis of a man who could make such a request. Lucian, from whom Apuleius derived most of the material for his famous romance, imagines himself in one of his satirical dialogues undergoing a like transformation through the conjurations of a Thessalian sorceress; and Boëthius, in his Consolation of Philosophy, describes similar metamorphoses: "One man," he says, "takes the form of a boar, another that of a Marmarian lion, others become howling wolves and fierce tigers."

Appuleius himself was a Platonist with a strong tendency to mysticism, and had the reputation of being a powerful magician. He was once brought to trial on the charge of having exercised this occult and uncanny faculty in what the Scotch would call an exceedingly canny way, by "enchanting" a rich widow and inducing her to marry him, to the great detriment and intense disgust of the heirs presumptive, who instituted a suit for damages. The court decided that, however irresistible may have been the influence he exerted in winning the affections of the lady, there was no ground for supposing that he had resorted to wizardry or any forbidden form of fascination. Romantic sentiment, in the opinion of the learned judge, sufficed to account for the attachment without the intervention of necromantic arts.

Perhaps the charm might have been explained, as Goethe, in a letter to Wieland, was fain to account for the ascendency gained over him by Frau von Stein, on the theory of metempsychosis: "Yes," he exclaims, "we were formerly man and wife." If this be so, one may naturally wonder what fatal act it was of his in that previous life that prevented the renewal of this pre-existent union and condemned him here to the conjugal care and companionship of Christine Vulpius.

A case quite similar to that recorded of Apuleius occurred in December, 1894, at Munich, Bavaria, where a Polish adventurer and itinerant practitioner of hypnotism and magnetism was accused, as the indictment runs, of having "inspired a lady with irresistible love through post-hypnotic suggestion in hypnotic sleep, and thus enabled himself to enter into the most intimate relations with her, and then by the same means deprived her of all recollection of what had taken place, thus causing her to deny that she had ever been hypnotized by him." The lady was Baroness von Zedlitz, a wealthy spinster of thirty-eight, whose property the defendant, Czeslav Czynski, who had a wife still living, attempted to get into his possession by a sham marriage, at which a certain Stanislaus Wartalski officiated as priest under the name of Simon Werthemann, D. D., signing and sealing the marriage certificate and carrying the feint and fraud so far as to toast the newly wedded couple as "duke" and "duchess" at the wedding dinner, which immediately followed the ceremony. The trial lasted several days and ended in the condemnation of Czynski to three years' imprisonment and five years' infamy, not, however, on the main accusation, of which the jury acquitted him, but on the collateral charges of instigation to an offence against public order by the assumption of a public office and the use of a forged public document.

It is interesting and instructive to note how extensively the idea of metempsychosis permeates and impregnates popular tales and superstitions. Mythology and folklore are full of stories of such transformations and transmigrations, the human soul entering into a flower, a shrub, a tree, a butterfly, a bird, a beast, or a reptile. In the Song of Roncesvalles a blackthorn is said to spring from the body of each painim who fell in battle, and a white flower of heavenly perfume from the head of every Christian warrior slain. The belief that the souls of the dead inhabit various plants, especially the rose, the lily, the linden, and the elder, is as old and widely diffused as the Aryan race.

Zoölatry or animal worship stands in intimate relation to metempsychosis. The primitive man was puzzled by the mysterious origin and nature of the lower animals and by the equally mysterious phenomenon of death and by the thoughts to which this event gave rise touching the departure and destiny of the soul. These two great mysteries were made to explain each other, the spirit of the man passing into the body of the animal, whose chief qualities he shared and to which he would, therefore, be drawn by the strongest ties of affinity. Beasts of prey, which feed on human flesh, were especial objects of worship, because they were supposed to be habitations of the spirits of the persons whom they had devoured, and it was deemed desirable to propitiate and appease these angry ghosts.

The Dakota Indian eats the liver of the dog in order to acquire the fleetness, courage, and hunting sagacity of this animal. Cannibalism, wherever it exists as an established custom or tribal institution, and not merely as a temporary refuge from stress of famine, originated in the conception of the possibility of transferring the spiritual attributes of animals and persons to those who consume their bodies and thus make them a part of themselves. The soul is not squeamish and migrates readily from one organism to another through the stomach.

A vague feeling of awe and mental awkwardness is awakened by the thought of a vagrant disembodied spirit; and the crude and crass imagination of the savage, not knowing what else to do with it, puts it into that other enigmatical incarnation of life, the beast. This is the lowest and grossest form of metempsychosis or metasomatosis. The beast, in which the human soul is thus re-embodied, becomes thereby an object of peculiar fear and reverence, and is supposed to be endowed with a certain mystical and supernatural power of doing good or evil, aside from the infliction of physical harm. The transmigration, in such cases, was not regarded as a punishment or degradation, but rather as a promotion to a higher plane of existence, a sort of apotheosis and deification. The tendency of the primitive man was to look upon the wild beasts of the chase, not as inferior, but as superior beings, whose force and faculty he viewed with envy, and to which he paid a ceremonial homage even in the act of killing them. The gods of rude peoples are, for the most part, zoömorphic, revealing themselves in brute forms. The natives of Africa adore the elephant, the hyena, and the crocodile, but have no representatives of the anthropoid race, such as apes and gorillas, in their pantheon; and the North American aborigines render divine honours to the owl, the beaver, the eagle, the bear, and the rattlesnake, but in no instance are their deities anthropomorphic. Survivals of this belief are found in the religions of the most highly civilized peoples. Even in Christianity the third person of the Trinity is symbolized as a dove. When a European appeared for the first time with an ass among the Quaquas, they at once made a god of the long-eared and loud-braying brute, watching every movement as ominous, and interpreting every harsh hee-haw as the voice of an oracle. The ass was a greater mystery to them than the white man, and they obeyed the universal law which governs the expression of religious feeling by prostrating themselves before it. The first impulse of the primitive man is to regard any strange creature as the embodiment of an evil spirit, a demonic incarnation, that may do him harm if not properly propitiated. The same feeling, directed toward inanimate objects, gives rise to hylozoism as an ontological theory and to fetichism as a religious cult.

With the assignment of the beast to its proper place in the order of evolution and the recognition of it as a creature lower than man in the scale of being, but having a genetic connection with him, the doctrine of the transmigration of souls acquired a deeper purpose and significance, less as a means of punishment arbitrarily applied than as a process of spiritual growth and progressive transformation. According to this more philosophical modification of the original theory, every living creature in the vast and complicated system of Nature is the embodiment of certain passions and affections suited to its degree of development, and every individual passes at death into the bodily organism of the animal which he has striven most assiduously and persistently to imitate while he was still humanly incarnated. Swedenborg states that one day, after having eaten more heartily than usual, he perceived a sort of vapour issuing from the pores of his skin and filling the room, until at length it began to descend and turned into hideous reptiles as soon as it touched the floor. The apparition of snakes would lead one naturally to infer that the Swedish seer had partaken too freely of that sweetly delusive and exceedingly heady beverage known as Swedish punch. But whatever may have been the immediate cause of this startling vision, he regarded it as prophetic of what a man must necessarily come to by indulging low animal appetites and thus attaining through the rigorous and immutable law of cause and effect the goal of his conscious or unconscious aspirations.

Metempsychosis is only applied metaphor, or metaphor literally interpreted and practically insisted upon, as when we speak of a gluttonous, rapacious, tricky, cruel, or generally offensive person as a hog, a vulture, a fox, a tiger, or a skunk. Death merely releases the soul from corporeal restraints and enables it to seek a habitation better suited to the gratification of its cherished desires, in obedience to the law of spiritual affinity and attraction. Thus Nature is constantly endeavouring to rectify incongruities and to produce perfect harmony in all her works. The soul is everywhere the plastic and creative principle, which moulds the physical elements to its own ideal, as the poet Spenser says:

> For of the soul the body form doth take,
> For soul is form and doth the body make.

Plato tells us that the pure soul, when it is set free from the body, is drawn to what is pure, and the base soul to what is base, like seeking like. Thus each individual predetermines in the formation of his character his fate and future associations, working out his own salvation in a profounder and more philosophical sense than is commonly attached to these words in modern pulpit phraseology.

In the Institutes of Manu and Yâjñavalkya, fate (*daivam*) and human effort (*purushkâra*) are harmonized by resolving the former into the latter on the theory of the pre-existence of the soul, fate being only the natural sequence of past actions or of "deeds done in a former body." "The accomplishment of an act," says the Indian lawgiver, "depends upon fate and human exertion; but what is here called fate is manifestly the resultant of acts performed in a previous stage of existence. Some expect success from fate or from the inherent nature of the thing, from time or from human agency; others, of superior perception, seek it in the union of all these factors. For as with a single wheel there can be no progress of a chariot, so fate without human effort can not be carried into effect." (Yâjñavalkyadharma-Śastra, i, 348–350.)

According to this theory, the element of fore-ordination, so far as it enters into a man's character and controls his conduct, does not result from the arbitrary decree of a higher power, but is the natural and necessary outcome of a universal law, and springs directly from the operations of his own will, which is the source of all the forces that predetermine his career and shape his destiny. Modern science also tends more and more to confirm this view. To what fearful extent man is the helpless creature and melancholy victim of prenatal influences and external circumstances, and how largely his nature is subdued to his moral and physical environment is evident from the startling light which statistics has thrown upon the operation of his so-called free will and power of self-determination in relation to suicides, murders, accidental deaths, marriages, and other social phenomena. Whether or not the world be the play and jugglery of the Absolute, and

> . . . the great globe itself,
> Yea, all which it inherit.

be in reality only the "unsubstantial pageant," to which Prospero compares it, and man the helpless toy of destiny, it is certain that our criminal codes show a constantly increasing tendency to admit the extenuating force of circumstances in judging of human actions, and our schemes of philanthropy and reform, discarding in a great measure the old machinery of moral appeal and hortatory homily, are directed more and more to the counteraction of hereditary propensities and the improvement of the external conditions of human life as the most efficient means of eradicating vice, diminishing crime, and elevating mankind.

Plato also maintains, or at least suggests, that the puzzling problem of the origin of *a priori* notions, innate ideas, intuitions, axioms, necessary postulates, and universal affirmations of the reason, may be most easily and satisfactorily solved by regarding them as survivals of knowledge acquired in a previous state of existence, the winnowed and garnered fruits of prenatal experience. These inherent truths and intuitional perceptions thus constitute the sum of man's permanent and imprescriptible intellectual acquisitions prior to the present period of his incarnation, and represent, so to speak, the consolidated spiritual capital, which survives the dissolution of the body and with which he begins another and higher stage of embodiment. "The mind," says Spinoza, "can not be absolutely destroyed with the body, but somewhat of it remains which is eternal. . . . There are rare minds, of which the principal part is eternal; so that they have scarce anything to fear from death."

Lessing held that the possession of five senses is only peculiar to our present temporary state of being, and that in the past we may have had less and in the future may become endowed with more than five senses. Even now we know of persons who go in and out among us, eating, drinking, merry-making, and marrying like ordinary mortals, yet who claim to have won for themselves the faculty of conceiving and perceiving four dimensions. Is this a prophecy of what is in store for us all, the isolated and individual foreshadowings of a future four-dimensioned existence for the race? The old mystic Henry More recognised the existence of a fourth dimension which he called essential spissitude (*spissitudo essentiæ*), and regarded it as an attribute of spirits: *Ubicunque vel plures vel plus essentiæ in aliquo ubi continetur, quam quod amplitudinem hujus adæquat, ubi agnoscitur quarta hæc dimensio quam appello spissitudinem essentialem.* (Enchiridion Metaphysicum, pars i, cap. 28, § 7.)

Lessing's theory as set forth in his dissertation on the possibility of man's having more than five senses (Dass mehr als fünf Sinne für den Menschen sein können) is based upon the doctrine of evolution or the gradual development of man out of a lower and less perfect organism. The substance of his *a priori* and metaphysical argument is as follows: The soul is endowed with infinite powers of apprehension and conception, which it attains, not at once, but in an infinite succession of time. Nature never goes by leaps, but by steps, which are often so small and slow as to be almost imperceptible; and since Nature contains many substances and forces incognizable by any of the senses, which now serve the soul as physical organs for the acquisition of knowledge, it is necessary to assume that there will be future stages of existence in which the soul will have senses capable of perceiving all the substances and forces of Nature. With our present number of senses we are unable to perceive a great variety of objects which are either too small for us or too large, too near, too far, or too subtile, so that we are constantly hemmed in and hindered in our pursuit of knowledge by bodily limitations and imperfections, and thus only partially comprehend the real relations and qualities of things. But it would be irrational to suppose that the soul is destined to grope forever in fruitless search after that which, through the want of proper or sufficient organs, it is incompetent to grasp. "This system of mine," adds Lessing, "is certainly the oldest of all philosophical systems; since it is, in fact, no other than the system of metempsychosis, or the doctrine of the pre-existence and transmigration of the soul, which was not only a subject of speculation with Pythagoras and Plato, but also engaged the attention of Egyptians, Chaldeans, and Persians, and, indeed, of all the sages of the East. This circumstance ought to produce a prepossession in its favour; for, in matters of pure speculation, the first and oldest opinion is always the most probable, because it was at once suggested by common sense." It is also interesting to note that this theory of the possible genesis of additional senses through the striving of the soul after knowledge corresponds, in a remarkable manner, to one of the chief factors in the modern doctrine of evolution, as expounded and emphasized by the Neolamarckian school of scientists, namely, the originary influence of mental effort in producing new bodily organs and faculties and effecting important modification of species.

Lessing would have regarded the recent discovery of the so-called X-rays by Prof. Röntgen as an excellent illustration and quasi-confirmation of his views, for here we have to do with vibrations or undulations in ether akin to light, and yet wholly imperceptible to the human eye. Owing to the imperfection of our senses the existence of these occult forces in Nature, of whose mysterious and manifold workings we are just beginning to form a vague and extremely limited conception, has hitherto escaped the keenest scientific observation. Even now we perceive the marvellous and almost magical effects, but the cause is hidden from our sight.

In pantheism there is, strictly speaking, no place for independent finite beings, since the Infinite is everything. Even the postulated all-god of this system of religion is not a personality outside of the universe, but a power immanent in the life of every part of it and subject to its immutable laws. So powerful and pervasive is this tendency in Eastern thought, that not even the hard, narrow, and anthropopathic monotheism of the Arabian prophet has been able to resist its disintegrating and transforming influence. In India, the Vedânta, with its indigenous and exuberant growth of ages of exegetical and metaphysical speculation, has completely overrun and metamorphosed the gaunt body of alcoranic divinity; and in Persia a highly mystical and poetical sofism has grown up in the very bosom of Mohammedanism.

In the Orient, the two chief representatives of pantheism and atheism, as organized cults sacerdotally and ceremonially equipped and clothed, as it were, in full canonicals, are Brahmanism and Buddhism.

There is another class of Indian atheists who are

called *Śûnyavâdinaḥ* (i. e., affirmers of emptiness or nonexistence) and whose doctrines are most distinctly embodied in a didactic poem entitled Sûnîsâr (i. e., Śûnyasâra, the essence of emptiness). Unlike the Buddhists, instead of making nothing of this world they make everything of it. They despise all religious rites and are thoroughly materialistic and sensualistic in their ideas. Their generosity springs from selfishness and their altruism is a refined and far-sighted egotism. Like the Sadducean author of Ecclesiastes (iii, 19–22; iv, 2–6), they believe only in the present life, and in getting the greatest possible sum of pleasure out of it before they "all turn to dust again." "Take and enjoy the good things of the world, and give also to others their share, since thereby your own enjoyment is increased. . . . Men die and pass away like leaves on the trees; new ones shoot forth as the old decay. Fix not your heart upon a withered leaf, but seek the shade of the green foliage. The horse that cost a thousand rupees, when dead is worthless, but the live nag bears you on your way. Trust not in the dead, but in the living; for he that is dead will never be alive again. This is a truth which all men know: of all those that have died not one has come back again or brought tidings of the rest. . . . The living care not for heaven or hell, and when the body is turned to dust what distinction is there between an ass and an ascetic?"

Curiously enough, the highest good or supreme bliss, which is the aim and aspiration of the mighty opposites, Brahmanism and Buddhism, is the same, namely, final and eternal exemption from the painful process of transmigration through the extinction of personal existence: the Brahman looking forward with cheerful hope to absorption in the universal spirit, and the Buddhist striving by suppressing evil passions and by seeking the "path of the law" or of duty (*dhammapadam*) to render himself worthy of attaining individual annihilation and of passing into the sinless and endless tranquility of Nirvâna.

Egotism is the essence of individuality, and in egotism every form of evil has its root. Neither austerities, nor ritual observances, nor almsgiving, nor good works of any kind have power to purge human nature of this universal taint. The only radical cure of self-love and self-assertion, the pride and naughtiness of the heart, is the utter extinction of individual existence, since these qualities are inherent in selfhood. The Apostle Paul says, "Though I bestow all my goods to feed the poor, and though I give my body to be burned, and have not charity, it profiteth me nothing." But the Sikh prophet Nanak declares, "Though I give my body as an offering to the fire, or cause it to be sawn asunder, or let it perish in the Himâlaya, yet will the malady still cling to my mind; and though I bestow in charity castles of gold and excellent horses and elephants and land and much cattle, yet will egotism abide within me." That "respect unto the recompense of the reward," which Paul praises in the conduct of Moses as one of the fruits of faith, is denounced by Nanak as a product of egotism. "Spotless," he says, "is the religion of that man who worketh and looketh not to the future reward."

According to this doctrine, the man who restrains his passions and rejects the pleasures of sense, leading a holy, virtuous, and beneficent life, with no hope or desire of personal remuneration either in this world or in the world to come, acts from higher and purer motives, and gives freer scope to the development of a morality untainted by selfishness, than he who consoles himself with the belief that his self-denial here will be compensated for by a thousandfold greater positive happiness hereafter, the light affliction, which is but for a moment, working for him a far more exceeding and eternal weight of glory. Here surely is more room for a system of moral duties than in a religion that offers to its votaries the allurements of a Christian heaven or a Mohammedan paradise as a reward for right conduct. The Buddhist believes that the effects of his good deeds are not dissipated by his death, but that, although he may cease to exist as a personal entity, his virtues live after him, entering into and increasing the moral inheritance of the race, perfecting, magnifying, and glorifying the great being—*le Grand Être*—humanity, easing the burdens of life for others, and mitigating the common misery of the sentient world long after his own individual consciousness has found the blissful end of all its strivings and wanderings in the eternal rest and peace of Nirvâna. "Do unto others as ye would that others should do unto you" is the golden rule; but far purer and more precious than gold is the injunction to do good without any reference to self, and to cultivate a morality that does not reflect the faintest tint, nor involve the slightest implication of self-love.

The alleged defect in the pantheistic scheme of salvation is the difficulty of harmonizing the decrees of fate, which are written on every man's forehead, with the constant appeals which are made to him as a free agent, accountable for his deeds. But what system of theology *has* ever succeeded in reconciling the sharp antitheses of fore-ordination and predetermination with personal responsibility? Surely not Milton's Stygian group of metaphysicians, who

> . . . reasoned high
> Of Providence, foreknowledge, will, and fate,
> Fixed fate, free will, foreknowledge absolute,
> And found no end, in wandering mazes lost;

nor the Greek poets and philosophers, with their notions of the influence of μοῖρα upon the destinies of gods and men; nor Paul with his theory of "the election of grace"; nor Augustine and Calvin with their dogma of the arbitrary predestination of men to eternal happiness or endless woe, a dogma which even the stern Genevan himself admitted to be a *doctrina horribilis*. Supralapsarian and Infralapsarian, Pelagian and Semi-Pelagian, the objective necessity of Thomas Aquinas, and the subjective necessity of Duns Scotus, after all their hair-splitting and logic-chopping, come no nearer to a satisfactory solution of the puzzling problem, and appear even more inconsistent and inconsequent in their reasonings about it than the Hindu pantheist, who regards all finite beings as mere modes and manifestations of the Supreme Being, just as waves are but fleeting forms of water, rising out of and remerging into the sea. Indeed, as we have already seen, the pantheistic identification of fate with the exercise of free will in a former state of existence comes nearer to a reconciliation of these conflicting forces than any other system of metaphysics or scheme of theodicy.

Again, from an ethical point of view, the inequalities of human conditions, the seemingly capricious distribution of good and evil, the pleasures enjoyed and the pains endured by men independently of any obvious relation to their respective characters or acknowledged deserts, can be best explained on this hypothesis, which, unlike the current orthodox theodicy, is at least competent to

> . . . assert eternal Providence,
> And justify the ways of God to men,

without diabolizing the Divine Being and utterly subverting our common conceptions of justice. It is surely more moral, as well as more intelligible, to suppose that the ills we suffer in this world are due to our own individual antecedent sins and shortcomings than that they are attributable to the transgressions of one far-off, reputed progenitor and federal head, for whose conduct no subtilty of casuistry can make us feel in the slightest degree responsible, or that our eternal destiny is determined by the arbitrary decree of a being,

> Wha, as it pleases best hissel',
> Sends ane to heaven and ten to hell,
> A' for his glory.

The proofs of personal immortality derived from the emotions or from the principle of compensation and retribution may be urged with equal cogency in support of transmigration. For this theory puts it into the power of every human being, and indeed of every living creature, to determine what form and feature the future life shall assume. Man is the maker of his own destiny in more than the proverbial sense of the phrase, elevating or degrading himself in the scale of sentient existence by his own acts. Each new incarnation that awaits him, like the maiden of heavenly beauty or hideous aspect, who meets the soul of the Parsi at the Chinvad bridge, is the personification of his own thoughts, words, and deeds. He grows into the complete embodiment of the propensities which he fosters, and fondly cherished tendencies take root in him as instincts, until by imperceptible gradations his

> . . . nature is subdued
> To what it works in, like the dyer's hand.

The recent progress of the physical sciences has also lent additional interest and importance, not to say probability, to the ancient doctrine. Metempsychosis would seem to be the spiritual counterpart of metamorphosis, the transmigration of souls being logically and analogically suggested as a corollary to the transmutation of species. The one does not necessarily involve the other, but both lie in the same line of thought. There is, furthermore, no reason why the theory of the conservation and persistence of force should not be applicable to mental or psychical, as well as to mechanical or physical forces. No impulse ever ceases, no motion is ever lost, no atom can be disturbed without disturbing every atom in the universe. If a sparrow fall to the ground, the momentum of its falling body is imparted to and affects every particle of the globe. But what becomes of the vital force which animated the bird and impelled it through the air?

It is, furthermore, an axiom of science that force is an essential attribute of matter. "Both," we are told, "are mutable, both indestructible, and both, so far as we know, quite incapable of existing alone." Mental operations are dependent upon material processes, which modern physiologists have succeeded, to some extent at least, in tracing. Of the one apart from the other we have no experience, and therefore no knowledge. Force isolated from matter and matter devoid of force are alike inconceivable. The materialist, then, should not, and in fact does not, deny the existence, but, on the contrary, emphatically asserts the eternity of spiritual force: he denies only the possibility of its existence, except as inhering in some material form, some solid, liquid, gaseous, visible or invisible, palpable or impalpable, ponderable or imponderable body. There would seem, therefore, to be little or no difference between the mere declaration of the immortality of the soul and the affirmation of the indestructibility of psychic force. The only question in dispute is as to the conditions under which this sentient principle, this thinking and conscious energy, survives and continues to operate. Does spirit remain forever a distinct personal entity, disembodied or reembodied, or is it, too, convertible into other forces, manifesting itself in manifold and interchangeable forms, like light, heat, magnetism, electricity, motion, and gravity? Science clearly indicates the latter; faith and the cravings of human affections cling to the former.

But whatever may be the nature and essence of the *scintilla animæ divinæ*, and whatever transformations it may undergo, we have never known it and can not conceive of it, except in connection with some more or less highly organized collocation of material atoms. This is true of both " celestial bodies and bodies terrestrial." The apostle can not describe them nor the imagination picture them otherwise than as substance, differing only in degrees of refinement and subtilty, as the sun differs from the moon or "star differs from star in glory." We can not, even in thought, disassociate force from something forcible, nor imagine wisdom or virtue as existing apart from the wise or the virtuous. Not only in actuality, but also in ideation, the abstract presents itself to us always and everywhere as the concrete.

From a purely speculative standpoint, therefore, the assumption that soul force is essentially distinct from all other forces, never being converted into any of them, but always preserving its individuality as a thinking entity, and the scientific axiom that all force is indestructible and inseparable from matter, would, when taken together and logically formulated, lead inevitably to the doctrine of metempsychosis. The beast soul has been characterized by Thomas Aquinas as *substantia incompleta ratione subsistentiæ et naturæ*, i. e., an entity which exists only as aided and supplemented by matter; but it is a question whether this dependence upon matter does not hold true of all souls.

We have no knowledge or experience of any force as an entity, but only as a phenomenon. We recognise it solely in some physical manifestation. What we call metaphorically the flame of life, the vital spark, may be the result of a combustion of gases, like any other flame, and when this chemical action ceases the flame goes out. The flame has, in fact, no real, but only a phenomenal existence; it is the visible effect of a process. Death is brought about by the operation of the same forces that produce and sustain life. There is nothing that leaves a person's body at death that has not been leaving it ever since his birth; only the loss, so to speak, is greater than the supply. The change is one which has always been going on, but in different degrees and relations. The debit begins to exceed the credit in the ledger of life; the balance keeps on fatally accumulating on the wrong page; the organism becomes painfully conscious of a deficit, until all further transactions are impossible, vital processes cease, and bankruptcy is inevitable.

It is not our present purpose to discuss the theory of the transmigration of souls as a tenet of philosophy, but merely to call attention to its beneficent influence as a code of morals and especially to its effect upon the relation of man to the lower animals and his kind and considerate treatment of them. The recognition of an original affinity between man and beast, however remote the kinship may be, or whether it be based upon the ancient dogma of metempsychosis or the modern doctrine of evolution, necessarily creates a current of sympathy extending even to the most insignificant members of the great and widely diversified family of sentient beings, and rendering it impossible willfully to neglect or maltreat the "poor relations," to whom we are united by the warm and living ties of blood.

In a few bold lines already quoted and written half a century ago Emerson anticipates the most radical deductions from Darwinism in his poetic conception of how,

>. . . striving to be man, the worm
>Mounts through all the spires of form.

A clear perception and abiding consciousness of this truth would cause even the most heedless wayfarer to take heed to his feet and step aside, rather than tread upon the humble embodiment of such lofty aspirations.

It would be unreasonable to suppose that no cruelty to animals occurs in Oriental countries, where metempsychosis is the prevailing speculative opinion. The observations of Ernst Häckel (Indische Reisebriefe), W. Heine (Eine Weltreise), Graul (Reise nach Ostindien), and J. Lockwood Kipling (Man and Beast in India) suffice to dissipate any illusion of this kind. Unfortunately the conduct of men is not always consistent with the religious precepts and philosophical principles by which they profess to be governed, and the strictest injunctions to kindness and compassion are often of little avail in resisting the primitive instincts and impulses of brutality inherent in human nature. The absolute prohibition of the destruction of animals, prescribed by Buddhism and Jainaism, is especially absurd in India, where savage beasts and venomous reptiles abound and put the inhabitants in daily peril of their lives. Far more sensible in theory, as well as more salutary in practice, is the discrimination of the Parsi between useful animals, creations of the beneficent spirit, which are to be carefully cherished, and noxious animals, creations of the hurtful spirit, which are to be conscientiously exterminated. The religious duty of preservation or destruction is in each case equally imperative.

But, notwithstanding its marked deficiencies and manifold disadvantages, the doctrine in question has undoubtedly produced in the East a tenderer regard for the rights of domestic and wild animals than is generally prevalent in the West, where, until quite recently, only such beasts and birds as were the property of man were thought to be entitled to any human sympathy or legal protection whatever. Cruelty was prohibited and punished solely as an infringement of the rights of the owner, for which the imposition of a fine would fully compensate him, but no consideration was given to the sufferings of the animal itself, which was regarded merely as an animated and automatic machine. It is also a significant circumstance that the earliest European advocates of animals' rights based their arguments and appeals upon panpsychism, or the essential unity of all forms of sentient existence, and upon the assumption that beasts are, like men, emanations from the infinite source of being and parts of the general soul of the universe. This is true of the Ionic school, the oldest group of Greek philosophers, of whom Anaximander anticipated the latest inferences from the doctrine of evolution concerning the descent of man, of Pythagoras, Empedocles, Theophrastus, the Stoics, Plotinus and Porphyry, and the Neoplatonists and Neopythagoreans in general. In modern times the same theory has been held by Hamann, Herder, Schleiermacher, Krause, Schopenhauer, Eduard von Hartmann, Lotze, Wundt, Paulsen, the materialists Ludwig Feuerbach, Moleschott, and Büchner, not to mention many less noted writers, and, so far as it affects the ethical relations of man to the lower animals, is elucidated and confirmed by the newest developments of biology and zoölogy, which began with the publication of Darwin's Origin of Species. With the disappearance of the crude hypothesis of special creations from the domain of natural history, the anthropocentric conception of the universe has ceased to be tenable and has been abandoned by the majority of scholars in every field of investigation, including even the best thinkers in the province of theology. The influence of this change of view in enlarging the scope of ethical inquiry so as to bring not only the lower races of mankind, but also the lower animals within its range, is especially observable, and has been most fruitful of happy results by awakening feelings of compassion and a sense of justice in individual minds and giving expression to them in municipal and national legislation.

A striking manifestation of this newly awakened sympathy is the organization and legal recognition of societies for the prevention of cruelty to animals. These benign and strictly secular institutions are of comparatively recent origin, and furnish, by the sad necessity of their existence, a confession and confirmation of a radical deficiency in Christian teaching, which they endeavour to supply. Such associations would be superfluous in Brahmanical or Buddhistic lands, where men are taught from their infancy to hold all life inviolably sacred, and kind and sympathetic treatment of the lower animals constitutes an essential element of religion and religious education. It is true that in every country and every community there are persons who are wholly unamenable to such instruction or to any sort of moral suasion, and on whom ethical teaching can be inculcated only by judicial punishment. In the conduct of life their sole criterion is the criminal code; whatever it prohibits and punishes they regard as wrong, and whatever it permits they assume to be right. The perfect man is, in their eyes, one who has never been guilty of a misdemeanour, for which he could be fined or sent to prison. Upon this class of individuals, which is much larger than it is generally supposed to be, penal legislation exerts an educational influence, serving as a permanent preventive of crime by elevating the average standard of public morality rather than as a temporary deterrent by appealing to the principle of fear. Thus lawgivers and courts of justice exercise correctional functions in a moral and didactic as well as in a purely punitive sense of the term; and it is especially in the extension of the sphere of criminal jurisprudence to the protection of animals against the capricious cruelty of man that its ethical value, as a means of moulding popular sentiment and moralizing public opinion, has been most perceptible.

In India hospitals for diseased and decrepit beasts have existed from time immemorial, and still constitute a universally recognised object of public charity and private munificence. Thus we find established in Bombay a flourishing institution of this kind, known as Panjara Pol, founded and supported by wealthy Jaina merchants and other Hindu sects, especially the votaries of Vishnu. It is richly endowed and situated in a street outside of the fort and covers several acres of ground. Prof. Monier Williams, who visited it, says: "'The animals are well fed and well tended, though it certainly seemed to me that the great majority would be more mercifully provided for by the application of a loaded pistol to their heads." This remark is doubtless correct, and would apply with equal force and pertinency to many suffering and incurably diseased persons. Savage tribes are wont to give expression to their compassionate feelings in this summary and effective manner. To the Greeks and other nations of antiquity it seemed as absurd to prolong the life of a decrepit man as it does to Prof. Williams to prolong the life of a decrepit beast. But a sense of the sacredness of human life prevents Englishmen of to-day from showing kindness to the aged and infirm by killing them, and a still stronger feeling of the same kind prevents Jainas from treating old and sickly animals in the same way. The difference consists merely in a narrower or broader application of the *Ahiñsâ* commandment: Thou shalt not kill. " A large proportion of space," continues our informant, " was allotted to stalls for sick and infirm oxen, some with bandaged eyes, some with crippled legs, some wrapped up in blankets and lying on straw beds. One huge, bloated, broken-down old bull in the last stage of decrepitude and disease was a pitiable object to behold. Then I noticed in other parts of the building singular specimens of emaciated buffaloes, limping horses, mangy dogs, apoplectic pigs, paralytic donkeys, featherless vultures, melancholy monkeys, comatose tortoises, besides a strange medley of cats, rats, and mice, small birds, reptiles, and even insects, in every stage of suffering and disease. In one corner a crane, with a kind of wooden leg, appeared to have spirit enough left to strut in a stately manner among a number of dolorous-looking ducks and depressed fowls. The most spiteful animals seemed to be tamed by their sufferings and the care they received. All were being tended, nursed, physicked, and fed, as if it were a sacred duty to prolong the existence of every living creature to the utmost possible extent. It is even said that men are paid to sleep on dirty woollen beds in different parts of the building, that the loathsome vermin with which they are infested may be supplied with their nightly meal of human blood." This last statement is doubtless the invention of some itinerant wag hailing from the home of Douglas Jerrold or the land of Mark Twain, although it must be confessed that the Oriental has the courage of his opinions very strongly developed and seldom shrinks from the logical application of his principles, no matter to what extremities they may reduce him. The first of the five commandments, which constitute the moral code of the Jainas and correspond almost exactly to the *pânchasîla* of the Buddhists, inculcates a tender regard for all forms of life. That this noble feeling should be carried to ridiculous excess and impose a number of absurd prescriptions, such as to strain water before drinking it, never to eat or drink anything in the dark, lest an insect might be inadvertently swallowed, to sweep the ground with a soft brush before sitting down lest an insect might be crushed, not to walk in the wind without wearing a piece of muslin over the mouth lest an insect might be blown into it, not to leave a liquid uncovered lest an insect might be drowned—these and many other equally preposterous precautions may travesty but can not destroy the beauty and worth of the fundamental idea. One can imagine the depths of horror and despair to which a conscientious Jaina would be consigned by a microscopic examination of his daily food and drink.

The feelings with which a visit to the Panjara Pol inspired the Oxford professor do not differ essentially from those which would be excited in any refined and sensitive mind by a walk through the wards of an ordinary hospital. We have already left far behind us the primitive barbarism, which would have laughed at our anxiety to promote the comfort and to preserve the lives of old and useless persons as foolish sentimentalism. Perhaps, when we have fully outgrown our anthropocentric ideas and traditions, we may also discover in a hospital for old and worn-out animals something really commendable and not utterly and irredeemably comical.

The Italian physiologist Prof. Mantegazza, in the record of his travels in India, describes a similar establishment, and seems to have been greatly disgusted with what he witnessed. One could hardly expect that such tender regard for subhuman infirmities would excite any other than loathsome feelings in the mind of the man who invented a new kind of rack, called the "tormentor," for the express purpose of inflicting upon animals the most excruciating pain of which he could possibly conceive. Day after day and month after month he contemplated, as he confesses, "with much delight and extreme patience" (con molto amore et pazienza moltissima), the sufferings of dogs and other exceedingly sensitive creatures stretched upon his horrid engine and enduring prolonged agonies, which, if we may judge from the meagre scientific results in his publications, served no purpose whatever except the gratification of a morbid and insensate curiosity.

On the 5th of March, 1890, there died in Bombay, at an advanced age, a Parsi woman, named Lady Sakarbai, whose husband had appropriated five years before a considerable sum of money to found a hospital for animals, which he called "Bai Sakarbai" in honour of his wife, and which was to remain a monument to her memory. During the rest of her life the now deceased lady took a lively interest in this philozoic foundation and left it in a flourishing condition.

Institutions of the kind just described are both charitable and educational. The compassion manifested in such cases not only alleviates the actual suffering of the beast, but it also exerts a wholesome reflex influence upon man, ennobling and humanizing his character, cultivating his affections and sympathies for the lower animals, and teaching children especially not to indulge in thoughtless cruelty toward any sentient creature.

When King Thibo, of Siam, sold a white elephant to Mr. Barnum, he stipulated in the contract or bill of sale that "the rich man who has bought the elephant agrees to love and cherish it, to make its life pleasant, and to keep it safe from all pain or injury." Tender consideration of this sort is a sentiment quite foreign to Christian civilization, and would be sneered at by the European, who does not scruple to send his broken-down horses to the knacker to be cut up into dog's meat, or to sell them for a song to a low and brutal carter to be driven and beaten to death in their old age.

Notwithstanding the ridicule which Prof. Williams heaps upon the Panjara Pol of Bombay, it has been deemed necessary to found a similar Animals' Institute in London, for the purpose of relieving the sufferings of sick or wounded animals by proper medical or surgical treatment. How urgent was the need of such an institution is evident from the fact that soon after it was opened for the reception of patients, the hospital was found insufficient to accommodate all the horses, dogs, cats, and other animals for which admission was sought. It was also thought advisable to establish, as supplementary to the hospital, a sanitarium in the suburbs of the city for convalescents and for cases requiring prolonged treatment, careful dietary, and rest. Although the animals of the poorer classes, as well as waifs and estrays, are treated gratuitously, the number of paying patients promises to make the institution self-supporting after the preliminary expenses have been covered.

Very different from this retreat for unfortunate animals is the veterinary hospital recently established in New York under the charge of four surgeons, the chief of whom also drives out to visit his patients in their homes like an ordinary medical practitioner. The principal patrons of this institution are wealthy ladies, whose pampered pugs, high-bred cats, and other pets suffer from indigestion caused by too rich and abundant food. Horses are frequently operated upon, but the cost of treatment is so great that it does not pay unless the animal has a value of several hundred dollars. From a general philozoic point of view this establishment has no practical value whatever, since it affords no relief to the thousands of maimed and sick creatures who stand in most pressing need of it.

In some of the other larger European cities we also find occasional asylums for stray and famished dogs and homes for houseless cats, such as the refuge founded by Ellen M. Gifford at Brighton, in England, for the succour and sustenance of needy animals, Miss Lindo's hospital for consumptive and home for weary horses near London, and the Countess De la Torres's asylum for cats at Hammersmith. The pound, which exists in most towns, or the parish pinfold, is an establishment of a wholly different nature, inasmuch as its purpose is not to provide a refuge for beast, but to give protection to man, corresponding, in this respect, not so much to a hospital or almshouse as to what the bedlam or the madhouse used to be before the discovery of more rational and scientific methods of treating the insane.

"In Egypt," says Lecky, "there are hospitals for superannuated cats, and the most loathsome insects are regarded with tenderness; but human life is treated as if it were of no account, and human suffering scarcely elicits a care. The same contrast appears more or less in all Eastern nations." Also some of the men most conspicuous for their activity during the Reign of Terror in France were very fond of pet animals. Couthon was strongly attached to a spaniel; Fournier lavished his love on a squirrel; Panis kept two gold pheasants; Chaumette had an aviary; and the sanguinary Marat was devoted to doves. The psychological problem presented in all these cases is to reconcile so much kindness to the lower animals with so great indifference or such excessive cruelty to human beings. It would be a mistake to suppose that the terrorists of the French Revolution did not love their fellow-men. On the contrary, so all-absorbing was their enthusiasm for humanity and so intense their affection for the race, that, as is often the case with philanthropists, they lost sight of the rights and were deaf to the woes of individuals. All other consideration were swallowed up in fanatical devotion to certain fixed ideas. We have an example of this perversion of feeling in readers of fiction, who waste their emotions in weeping over the trials and afflictions of imaginary personages and turn away with dry eyes and cold hearts from the misery of men and women in real life. The Oriental, whose extreme carefulness of beasts renders him careless of mankind, illustrates the same emotional limitations of human nature narrowed and intensified by religious superstition. This principle is exemplified on a smaller scale by the German lady who advertised in a Berlin paper for "well-mannered and well-dressed children to be employed for several hours each day to amuse a sickly cat"; and by the American lady who ordered a rosewood coffin lined with satin and inlaid with silver for the obsequies of a deceased lapdog. Peoples, as well as persons, may have their sympathies warped and drawn awry and thus develop into "cranks."

As a rule, in Occidental countries the first and prevailing impulse of the police authorities, as well as of the public in general, is to knock all stray and helpless animals on the head, and in most cases this summary method of proceeding is adopted. Mrs. Jameson gives the following account of what she once saw in Vienna at a time when there was a great dread of hydrophobia, and orders were issued to massacre all unclaimed or unmuzzled dogs found within the precincts or in the suburbs of the city. The men employed for this purpose were armed with a short heavy club, which they hurled at the proscribed animal with such force as to kill or cripple it at a single blow. "It happened one day that, close to the edge of the river, near the Ferdinand's Brücke, one of these men flung his stick at a wretched dog, but with such bad aim that it fell into the river. The poor animal, following its instincts or its teaching, immediately plunged in, redeemed the stick, and laid it down at the feet of its owner, who snatching it up, dashed out the creature's brains." And yet Christian legislation, the civilization which claims to be based on a religion of mercy and compassion, has no law to punish such a monster of cruelty and base ingratitude, but rewards him for his ignominious deed. In fact, the conduct of this vile fellow was only the logical outcome and rude application of our current anthropocentric ethics, as its effects are necessarily exhibited in a coarse and common nature. "A ruffian in the midst of Christendom," says Father Taylor, "is the savage of savages." But here the development of ruffianism is due directly to the influence of Christian zoöpsychology and the brutal exercise of what Shelley calls the "terrible prerogative" which it confers upon man.

It is for these reasons that charitable foundations for animals are usually regarded and often ridiculed as the amiable idiosyncrasies of eccentric individuals, or as the manifestations of a mild and harmless monomania peculiar to old maids and withered beldames, who, having found no worthier outlet for their loving natures, are content to pour the flood of their pent-up affections into this channel. It is, in sooth, a curious circumstance, and quite significant of the character of our civilization, that endowments of this kind are not with us, as in the East, the normal and legitimate expression of a humane and benevolent spirit, but rather serve incidentally as the waste pipe of suppressed and soured emotions, having their real source in a generous and sensitive nature perverted by pessimistic and misanthropic views of life. Thence it comes that, with us Occidentals, the love of animals, instead of being the proper expansion of philanthropic sentiment, too often springs directly from intense and morbid hatred of mankind. It was this feeling that made Schopenhauer shun the society of his fellow-men during life, and in dying bequeath his property to his poodles. "Men," he declared, "are the devils of the earth, and animals are the souls which they take pleasure in tormenting. This state of things is the consequence of that installation scene in the garden of Eden." Solomon, too, in one of his spleeny and cynical moods, when he "hated life," and, surfeited with its pleasures, denounced them all as "vanity and vexation of spirit," affirmed that "a man hath no pre-eminence above a beast"; but in making this remark his purpose was not to elevate the beast, but to degrade the man. Both are reduced to the same plane of transitory existence by a process, not of levelling up, but of levelling down. "All go to one place, all are of the dust, and all turn to dust again. Who knoweth the spirit of man, whether it goeth upward, and the spirit of the beast, whether it goeth downward to the earth?" The implication of the passage is that however gross and grovelling a beast may be, man is no better.

In many portions of the East it is customary for Brahmans and Buddhists to express their joy and gratitude on recovering from sickness or on receiving any good fortune, not by chanting a *Te Deum*, but by going to the market place, where wild birds are exposed for sale by Jewish, Mohammedan, and Christian fowlers, purchasing a number of them, carrying them to the city gates, opening their cages, and restoring to the captives their former liberty. Under similar circumstances a European would most probably return thanks by inviting his friends to eat birds with him—just as the typical Englishman thinks the best use he can make of a fine day is to go out and kill something. There can be no doubt that this general attitude of mind is, in a great degree, the result of our current religious ideas and traditions and the early training that grows out of them. We are ourselves hardly conscious how deeply ingrained are our prejudices on this point, and how very difficult it is to escape the insensible pressure of these moral influences which inclose us like an atmosphere.

Not long ago a German Protestant parson, when asked to preach a sermon in support of a society for the prevention of cruelty to animals, replied that, although heartily sympathizing with the cause, he could not accede to the request, since the Bible did not furnish him with any text appropriate to such a discourse. As a humane man, he would be willing to make a speech in favour of it outside of the pulpit; but as a clergyman and divinely commissioned expounder of the sacred Scriptures, he was forced to pass it over in silence. Evidently the good parson was not well versed in the cunning arts of modern homiletics, and had little skill in the marvellous exegetic jugglery which easily conjures into passages of Holy Writ ideas and principles of which the writers never dreamed; otherwise he might have simply cut the Bible for his text, as was the practice of ancient sortilege, and preached from any passage thus selected a sermon suitable to the occasion.

Some years since the Thierschutzverein of Munich issued an appeal to the public, stating the aims and objects of the association, and seeking to rouse the lethargic Bavarians to a more earnest appreciation of its usefulness and to greater liberality in its behalf. In addition to purely secular considerations and motives of mere morality, the appeal was also urged on religious grounds, and sustained by the following quotations from the Bible: "A righteous man regardeth the life of his beast; but the tender mercies of the wicked are cruel" (Prov. xii, 10). "He giveth to the beast his food, and to the young ravens which cry" (Ps. cxlvii,

9). "He causeth the grass to grow for the cattle, and herb for the service of man: that he may bring forth food out of the earth" (Ps. civ. 14). "Who provideth for the raven his food? when his young ones cry unto God, they wander for lack of meat" (Job xxxviii, 41). "Are not two sparrows sold for a farthing? and one of them shall not fall on the ground without your Father" (Matt. x, 29).

There could be no better illustration of the poverty of our Holy Scriptures on this subject, and the little thought given to it by their authors than the citation of these texts, not one of which (except, perhaps, the first) has the slightest relevancy or was meant to teach kindness to animals, and to inculcate the principles advocated in the Munich circular. Even the passage from the Proverbs is a mere statement of fact, designed to illustrate the character of the righteous man, who regardeth even the life of his beast, and is contrasted with the wicked, whose bowels (as it ought to be translated) are cruel. There is no recognition of the rights of the beast, and no injunction to respect them. The whole reference is to man and the sense of his own worthiness as his standard of conduct. In the other verses, cattle, ravens, and sparrows are mentioned simply to show the watchful care and providence of God toward man. Here and there we meet with an isolated intimation of compensatory justice or tender feeling, as in the paragraph of the Mosaic law prohibiting the muzzling of the ox when it treadeth out the corn, and the sentimental or sanitary scruple about seething a kid in its mother's milk. It was the same crude and rudimentary conception of compensation that led the Greeks to decree that the asses which bore the stones for building the temple of Eleusis should be permitted to graze with impunity within the sacred grounds.

The Jews were also forbidden to take the parent bird while "sitting upon the young or upon the eggs," although they were permitted to rob her of the young. This provision was unquestionably a wise one, intended to prevent the reckless destruction and consequent diminution of the supply of birds. But there is no element of kindness or compassion in it, any more than there is in modern laws for the preservation of game, which are designed solely to insure and increase the pleasures of the chase, protecting animals in order to enhance the sport of hunting and killing them. The regulation was of a purely prudential and economical character, like that contained in the same code, forbidding the husbandman to sow divers seeds in his vineyard and thus deteriorate the quality of the grapes. The lawgiver was not moved by mercy to enact the former provision of the law any more than the latter.

In like manner William Cowper says:

> I would not enter on my list of friends
> (Though graced with polish'd manners and fine sense,
> Yet wanting sensibility) the man
> Who needlessly sets foot upon a worm.

But this assertion does not imply on the part of the poet any high appreciation of the worth of worms, such as Darwin shows in describing the important functions which they perform in the economy of Nature, but instances them on account of their supposed worthlessness, in order to emphasize his estimation of sensibility as an ornament of human character.

Throughout the Old and New Testament animals

are always regarded from an anthropocentric point of view, or in some satellitic relation to man. They are pronounced clean or unclean, not on account of their own habits and propensities, but according to an arbitrary standard of ceremonial purity, intended to secure his ritual or constructive cleanliness. He classifies them, like fungi, into edible and inedible, or "the beast that may be eaten and the beast that may not be eaten." They are made the scapegoats of his iniquities; and minute descriptions are given of their sacrificial qualities and uses, whereby their innocent and untainted blood is shed in expiation of human trespasses and sins. They are punished for his offences. Because the Israelities were incredulous and disobedient, God not only laid waste their vines and their sycamore trees, but "he gave up their cattle also to the hail, and their flocks to hot thunderbolts." Peter looks upon them merely as "natural brute beasts made to be taken and destroyed."

This spirit which everywhere prevails in the Jewish Scriptures and the Gospel records is predominant in patristic literature and mediæval hagiology. It is said of Cardinal Bellarmine that he used to let bugs and insects bite him undisturbed, on the plea that "we shall have heaven to reward us for our temporal sufferings, but these poor creatures have nothing to look forward to except the enjoyment of the present life." According to his fellow-Jesuit and biographer Fuligatti, his object, however, was not so much to gratify and regale the vermin, as to exercise his own patience and prepare his soul for paradise. "For this reason he would not brush away flies from his face, although they are wont to be very annoying, especially at Rome in summer." It was less an act of kindness than a lazy and nasty means of grace and of sanctification.

The same penitential path to holiness was pursued by St. Macarius, of whom an old chronicler relates: "It happed on a tyme that he kylled a flee that bote hym; and when he sawe the blode of this flee, he repented hym, and anone unclothed hym, and wente naked in the deserte vi. monethes and suffred hymselfe to be byten of flyes."

The lives of the saints are full of legends concerning their friendly and familiar relations with wild animals, and these stories are adduced as proofs of the power of holiness even over the brute creation. Thus the beasts and birds of the forest are said to have come at the call of St. Columbanus, flocking and frolicking about him like kittens (*ludentes laetitia, velut catuli*), and squirrels descended from the trees and sat on his shoulders or nestled in the folds of his mantle. A pack of wolves passed by him as he was kneeling in prayer, and did him no harm; and at his command a troublesome bear, which infested a valley near Anegray, quit the country and never returned. A mischievous raven stole his mittens (*tegumenta manum aut wantos*); but the saint threatened that all its callow young should die unless the mittens were immediately restored, which was done accordingly. St. Gall spoke to a bear in Latin, ordering it to bring a stick of wood for the fire, and Bruin, whose available knowledge of this language was evidently superior to that of many a modern professor, obeyed forthwith. St. Goar bade the hinds (*cervas*) come out of the wood and be milked, and, unlike the spirits which Glendower could call from the vasty deep, they came when they were summoned. The same holy man, on paying a visit to the bishop, hung his hat on a sunbeam which came in through the window. Whether the hat would still remain suspended in the air or fall to the floor, if the sun should chance to go under a cloud, is a point left undecided by the hagiologist.

Legendary literature records "a deal of skimble-skamble stuff" of this sort, which it would be tedious to repeat. All the glory of these acts haloes round the brows of the saints, whose kindly fellowship with the lower animals, although the natural result of a solitary, anchoretic life in regions remote from human habitations, is regarded as something miraculous, and has therefore failed to influence the conduct of ordinary mortals to any great extent. It is said that until the beginning of the seventeenth century it was deemed sacrilegious to kill a hare in the parish on the Tanat, in which was the shrine of St. Monacella, the protectress of hares. As a rule, however, such saintly guardianship contributes very little to the security of the creatures under their tutelary care.

St. James of Venice, a saint of the thirteenth century, used to buy and release the birds tied up and tortured by Italian boys, and Leonardi da Vinci was accustomed to purchase caged birds and set them free. It is also related of Pythagoras that he once bought the entire draught of a fisherman's net near Metapontus and restored the fish to their native element. But these isolated exhibitions of tender charity have not diminished the slaughter nor prevented the caging of small birds in Italy, nor have they saved the eyes of a single thrush from the hot iron with which the Italians are wont to destroy the sight of these songsters, in order that the perpetual darkness and loneliness of their lives may not only increase the quantity of their song, but also impart to it a peculiar quality and sweet strain of sadness.

St. Anthony of Padua preached the gospel to the fishes; but whatever effect his sermons may have had in saving them in the future life from the devil's toasting fork, they were of no avail in rescuing a single finny denizen of the deep from the frying pan in this world. The sole object of the story is to illustrate and glorify the moving eloquence of the saint, who, after delivering his homily, may have gone back to his cloister and dined on his parishioners from the pond with as much relish as a backsliding Fiji neophyte would enjoy a sparerib of his proselyter and pastor. Pious Christians and good Catholics do not deny themselves the exciting pleasures of deerstalking, because St. Hubert had a vision of the cross between the antlers of a stag, and gave himself up to a life of religious meditation; on the contrary, the canonized Bishop of Liège has become the patron of hunters and the protector of the chase. The legend of the wolf of Gubbio, which, at the injunction of St. Francis, abstained from mutton and obtained its food by going from house to house in the village, like a begging friar, has had no reformatory effect on wolves in general, nor was it intended to indicate a possible or desirable change of lupine habits, but solely to exhibit the power of the saintliness capable of working such a miracle. The practical ethical value of all these myths is simply null.

As the records of ecclesiastical excommunication show, it is only over noxious animals, for the purpose of cursing them, that the Church has claimed jurisdiction or cared to assert it. In behalf of the countless beasts which toil out their blameless lives in the service and at the mercy of man, she never utters a word of authority, nor lifts her crooked fingers in the form of benediction. True, she assigns a place in the calendar to St. Anthony, the patron and nominal protector of animals; and from the 17th to the 23d of January Romans of all classes—princes, peasants, cardinals, cabmen, and *campagnuoli*—used to bring their horses and asses to be blessed and sprinkled with holy water before the old church of S. Antonia Abbate in the Piazza di S. Maria Maggiore. But in most cases, where some merciful intervention was actually needed in favour of overworked and much-abused hacks and cart horses, this *festa* proved to be a holiday for the beast far less than for its owner, who rode through the streets arrayed in his best apparel, and adorned with feathers and ribbons of brilliant hues, and spent the day in careering from one wine shop to another and carousing with his friends. The ceremony, which always seemed to be performed perfunctorily, as though it were deemed an indulgent concession to the *sancta simplicitas* of the old Franciscan, was accepted as a joke and utilized as a lark, and never exerted any appreciable influence in restraining violence or inspiring kindness toward the lower animals. Indeed, the tutelar saint is seldom invoked during the rest of the year, except in maledictions. "May St. Anthony smite you!" is still the *popolano's* favourite imprecation on his horse or donkey; and fearing lest the request should not be granted, or willing to show his faith by his works, he plies the lash or wields the cruel *pungolo* in the saint's stead.

A similar festival is celebrated in the highlands of Bavaria and especially at Tölz, on the 6th of November, in honour of St. Leonhard, the local patron of horses and neat cattle, which on that occasion are curiously adorned with many-coloured fillets and flags and driven in procession, attended by priests with the sacred emblems of the altar and holy banners and all the cheap pomp and tinsel trappings of the Church. In this corso (*Leonhardfahrt*) each peasant strives to outdo the other in gaudiness of equipment, and thinks more of his own bravery than of the comfort of the quadrupeds for whose welfare the feast is supposed to have been instituted. Whatever benefit may accrue to the brute is purely incidental and wholly secondary to the pride and pleasure of the owner.

The mystic and visionary *pater seraphicus*, Francis of Assisi, sang his *Cantico delle Creature*, in which he thanked the Lord for brother sun, and sister moon, and mother earth prolific of fruits and flowers. He even greeted the wind and the fire as brothers, and the water and bodily death as sisters, but passed over, in significant silence, all sentient creatures and recognised no kinship with beast, or bird, or creeping thing. On another occasion, it is true, as he was taking a walk near Beragna, he addressed the birds as his "winged brothers," and bade them praise their Creator and love Him with all their heart. And the birds, it is added, came to him and perched on his hand and let him stroke their plumage and would not depart from him until he made the sign of the cross over them and dismissed them with his blessing. In the neighbourhood of Greccia he freed a hare from a snare and the grateful creature took refuge in his bosom and refused to leave him. Little lambs are also said to have followed after him, which, however, is by no means a marvellous thing for little lambs to do. He reproached a butcher, asking, "Why do you hang up and torture the lambs, my brothers, in this manner?" To this naïve and utterly idle question the butcher might have replied that his "sin's not accidental, but a trade," and that if it is a crime to slaughter sheep, then the eater of mutton chops must be regarded as *particeps criminis*. Again, he expressed his sympathy for some turtle doves in a cage, saying: "Why have you, my dear sisters, simple, innocent, and chaste creatures, allowed yourselves to be caught?"—a remark that would seem to censure the foolishness of the birds, rather than the ruthlessness of the fowler. Indeed, all this cheap commiseration of suffering creatures remained a barren sentiment, which did not contribute one jot or one tittle to the alleviation of their present distress, nor tend in the least to prevent its recurrence. There is no evidence that the example of the soft-hearted saints ever converted a single hard-hearted sinner from the error of his ways and led him henceforth to treat the lower animals with tenderer care and consideration. Nor is there any necessity of discarding these strange stories as mere pious fictions. Wild animals might easily be attracted to a gentle hermit in the solitude of the forest, particularly as they were, in most cases, unaccustomed to human beings and had not yet learned to fear them. Birds and beasts on islands uninhabited by man have uniformly shown the most perfect confidence in the discoverers of these islands and only learned by experience that man was to be avoided as their enemy. Some individuals have a certain magnetic influence over animals. Hawthorne ascribes this peculiar power to the faunlike Donatello, and Thoreau could put himself into relations of sympathy with the mute, cold-blooded, and unsocial fish and make it swim into his hand. King Ludwig I of Bavaria used to admire and envy an old woman, to whom the birds in the Court Garden at Munich would come in flocks, fluttering about her head and perching on her shoulders. His Majesty endeavoured to inspire them with the same confidence, but no calling or coaxing could induce them to approach him. He could not understand why they should prefer to light on a plebeian rather than on a royal hand; and finally in despair of their vulgar taste desisted from all further efforts to win their favour and settled a pension for life on the old hag who could work such witchery.

"Fromm waren die Münchener zu jeder Zeit," says one of their most quaint and genial chroniclers; and there are few cities in Europe where the priests are more zealous or exert a greater influence over the populace, or where the authority of the Church is more respected than in Munich. Yet no voice of warning or reproof was ever heard from chancel or confessional against the cruelty to animals, which used to disgrace the Bavarian capital. Veal is the favourite food of the inhabitants, and is consumed in enormous quantities; and it is no exaggeration to affirm that, before the erection of the slaughterhouse outside of the city, every calf carried to the shambles was made to suffer *in transitu* the tortures of a man crucified with his head downward. An archiepiscopal *Hirtenbrief* would have sufficed to check this brutality toward beasts; but unfortunately no such writ of mercy, no pastoral letter of pity, ever issued from the palace in *Promenadenstrasse*. The late pope, ninth of Piuses and first of infallible pontiffs, decided *ex-cathedra* that animals have no soul, and that, therefore, we are not bound to them by any of those moral duties and sacred obligations which we owe, in general, to our fellow-men, and in particular to them that are of the household of faith and are united to us by ties of religion. This opinion is fully indorsed and practically exemplified by the Italian donkey driver, who, to every remonstrance against the wanton beating and bruising of his patient beast of burden, retorts "*Non é cristiano,*" at the same time dealing a succession of vigorous thwacks with a heavy cudgel by way of adding emphasis to his dogmatic assertion. Thus, the bipedal brute continues to maul and maim the quadrupedal beast in the spirit of the Vaticanic dictum, and in accordance with a religious principle so clear and simple as to be comprehensible to the dullest *asinaio:* the poor creature is not a Christian, and therefore has no rights which a good Catholic is bound to respect.*

"*Non é cosa battezzata*" (it is not a baptized thing) is the Italian peasant's justification of any suffering he may wantonly inflict upon the lower forms of life. In England the cruel pastime of "cock-throwing," formerly practised by men and especially by schoolboys on Shrove Tuesday, has claimed a religious origin and consecration by being brought into causal connection with Peter's denial of Christ, as the poet Sedley sings:

---

* It is interesting to note that in Sheffield, England, "Christian" is popularly used to signify a man in distinction from a brute beast.

> Mayst thou be punished for St. Peter's crime,
> And on Shrove Tuesday perish in thy prime.

Since the crowing of the cock served as a rebuke to the recreant apostle and caused him to repent of his treachery, it is difficult to see why this valiant and vigilant fowl should be held responsible for his cowardice. On the contrary, one would imagine that all cocks would henceforth be as highly honoured and fondly cherished in Christendom as the descendants of the geese, which saved the Capitol, were by the ancient Romans. But it is the fatality of all vicarious schemes of retribution to reverse our natural and unperverted conceptions of justice and to make the innocent expiate the misdeeds of the guilty.

The efforts of some of the popes to suppress the Spanish bullfights were due, not to any pity for the tortured animals, but solely to the desire to prevent the destruction of human life; and these disgusting spectacles, which are the favourite sport of the most Christian nation of Europe, still take place under the auspices of the Church, a chapel, in which mass is read before the massacre begins, being connected with the arena. There the picador says his prayers and the functions of religion are most incongruously mixed up with *funciones de toros*. That the weekly proceeds of these holiday butcheries in Madrid should be devoted to the general hospital is a most striking example of anthropocentric selfishness and an unconscious satire on Christian charity. Indeed, when a society for the prevention of cruelty to animals was first established at Madrid, the Spaniards, whom foreign influences and fashion had brought into sympathy with the movement, proposed that a grand bullfight should be arranged in order to raise funds for the newly organized and merciful institution.

The idea of blood-relationship, which, as we have already shown, formed the basis of primitive society and of which the doctrine of humanity is but a wider development, has received still further extension through recent scientific researches tending to establish a genealogical connection between man and the lower animals. There can be no doubt that the general acceptance of the theory of evolution would exert upon the Western mind a wholesome influence in favour of greater consideration for all forms and embodiments of life, corresponding to the benign effect which the belief in metempsychosis has produced upon the less positive and more mystical and metaphysical mind of the East.

> Wer sich selbst und Andre kennt
>   Wird auch hier erkennen;
> Orient und Occident
>   Sind nicht mehr zu trennen.

> Who knows his own and others' bent
>   Will here, too, clearly see
> That Orient and Occident
>   Can no more severed be.

Not only is the present drift of scientific research strongly set in this direction, but minds of the highest culture in every department of thought share in the same movement. Thus comparative psychology, as will be shown in a subsequent chapter, is gradually overturning one barrier after another, which a narrow and obsolescent metaphysics had erected between man and beast in respect to their mental faculties and moral qualities, and even the comparative study of languages is rapidly removing from this field of investigation artificial obstacles of a like character, which an antiquated philology had declared to be fixed and impassable. These points are fully discussed elsewhere and are referred to here only to indicate their ethical bearings upon the question of animals' rights and to show the practical agreement, in this respect, between the results of Oriental speculation and of modern evolutionary science.

The metempsychosist holds that the *scintilla animæ divinæ* or divine spark wanders through eight million four hundred thousand creatures before it is fit to animate a human being. Still every incarnation is an essential and sacred link in the unbroken chain of existence that connects the mollusk with man and slowly lifts the whole out of the mirage of phenomena and the illusions of selfhood to ultimate reunion with the Supreme and Eternal Spirit, from which it emanated, and which is the only reality. The evolutionist teaches that the struggle for existence and the survival of the fittest through natural selection went on millions of years through successive ages before the principle of life or intelligence found its highest embodiment in man. According to both of these theories man is not an isolated product of Nature, called into existence by a divine fiat, but a part of the general order of things with no break in the continuity of his development out of the lowest organisms from the protoplasmic cell upwards.

It is evident that our moral and religious instruction, based upon the anthropocentric assumptions of Judaism and Christianity, has been hitherto lamentably defective. Perhaps, with the introduction of more rational views of cosmogony and anthropology, and broader and more generous principles of psychology into our elementary text-books, through the union of a sounder physics with a larger metaphysics, our children's children may finally learn that there are inalienable animal as well as human rights, and that, in respect to the ties of moral obligation and the claims to kind and just treatment which they imply, not only "all nations of men," as Paul affirmed on Mar's Hill, but, as the Indian sage declared, "all living creatures are of one blood." To the Hebrew decalogue and the Christian beatitudes must be added the first of Buddha's ten commandments:

> Kill not for Pity's sake, nor dare to slay
> The meanest creature on its upward way.

II.

ANIMAL PSYCHOLOGY.

CHAPTER V.

MIND IN MAN AND BRUTE.

Oriental speculation and Occidental science. Metempsychosis and evolution. Psychical kinship of man and brute. Automatic and volitional mental action. Freedom of the will in man and the lower animals. Consciousness in the lowest organisms. Protoplasm and protista. Chemical fabrication of products of vital forces. Berzelius and Wöhler. Schneider's classification of animal impulses. The nutritive impulse and its final purpose. Oken's classification. Impulses of sensation, perception, conception, and thought in the order of their development. Conjoint action of these impulses in the mental activity of the lower animals. Interesting experiment of Möbius with a pike. Incorrect inferences. How a horse learns the meaning of "whoa." Pains taken by parent birds to teach their young. Heine's lizard. Untenable distinctions between men and brutes. Too great importance attached to man's ability to look upward. Herbart's exact and concise statement of the grounds of man's superiority. Influence of infancy on human progress. Form and flexibility of the hand. Mental operations the spiritualizations of manual operations. Didactic value of mechanical labour.

IF we compare the latest achievements of Western thought with the results of Eastern speculation, we find in the doctrine of evolution a striking confirmation of the genetic and essential unity of organic nature, which the theory of metempsychosis assumes. Occidental science has firmly established what Oriental metaphysics only vaguely dreamed of. The seemingly fantastic and extravagant assertions of Indian sages concerning the transmigrations of the soul, and the countless ages of its successive reincarnations in its upward strivings toward the goal of complete emancipation from material existence, are but lengthened foreshadowings and grotesque adumbrations of the doctrine of natural selection and progressive development through the struggle for existence, involving perpetual adaptations to changes of environment that have been going on for millions of years, and producing organisms in which the intellectual faculty frees itself more and more from the bondage of material conditions, and asserts with constantly increasing emphasis its supremacy over mere brute force.*

Modern scientific research has not only discovered a multitude of physical correspondences—analogical and homological—between man and brute—but it has

---

* Kapila, the founder of the Sânkhya school of philosophy, may thus be regarded, in a certain sense, as the Indo-Aryan prototype of Darwin. The problems which they endeavour to solve are much the same, and their methods differ only as the poetic and mystic genius of the Hindu differs from the positive and matter-of-fact genius of the Englishman. In Kapila's writings, the Sânkhya Pravachana Sûtra and Sânkhya Kârikâ, there are many thoughts and expressions that would fit admirably into the Origin of Species. This famous muni discarded revelation and recognised no other final cause than great creating nature (*mûlaprakriti*); and his philosophical system is characterized by native scholiasts as *nirisvara*, usually translated "atheistic," but really signifying "agnostic."

also detected and brought to light many irrefragable proofs of their psychical kinship. The more exact and extended our knowledge of animal intelligence becomes, the more remarkable does its resemblance to human intelligence appear. The attempt to discriminate between them by referring all operations of the former to instinct and all operations of the latter to reason is now generally abandoned. Automatic mental action is known to characterize men far more, and the lower animals far less, than psychologists formerly supposed. In an hypnotic state the conscious psychical activities of the individual, as regards the exercise of his rational and volitional powers, are almost wholly suspended and superseded by automatic movements and alien impulses of suggestion, over which he has no control.

Indeed, there is strong presumptive evidence that *consciousness*, which is indicated by the simplest exercise of choice, and may be regarded as the distinctive peculiarity and fundamental element of Mind, manifests itself in the lowest forms of life, and is present even in protoplasmic and protozoic organisms. From this starting point the process of development is gradual, but continuous, from the *amœba* to man.

No psychologist has as yet been able to draw a hard and fast line between volitional, instinctive, and reflex actions, or to determine with any degree of precision what activities are attributable to each. It is highly probable that the so-called self-determinations of the will are as mechanical in their origin, and as definitely fixed in their operation under the influence of motives of various kinds, as are reflex actions under the influence of their appropriate stimuli. If we could trace all the complex incitements and impulses which lead the assassin to lift his arm and strike the fatal blow, we should doubtless find the necessity of the action as absolute and inevitable as the movement by which the decapitated frog raises its leg to scratch an irritative drop of nitric acid from its side. The argument in favour of human freedom, based upon an appeal to consciousness, has no validity whatever, since the forces, of which the act of willing is the resultant, lie outside of the sphere and beyond the cognizance of consciousness. Back of the mere recognition of the fact that an action is performed in obedience to the will—and this is as far as the power of consciousness extends or can claim any authority—is the profounder and more mysterious problem of the origin and constitution of the will itself, of which consciousness can have no immediate knowledge and furnish no satisfactory solution. What a man may will to do, when acted upon by certain inducements or temptations, was prearranged long before his birth, not by the arbitrary decree of a vindictive deity, but by prenatal influences and hereditary tendencies, facts of organization which may be subsequently modified by the social and moral environment into which he is born and the effects of early education. This is the truth which is symbolically expressed by the dogma of predestination, a *doctrina horribilis*, as Calvin himself admitted it to be, that loses nothing of this awful character by being transferred from the province of theology to that of physiology.

It is true that we perceive an immense disparity between the highest human and the lowest animal intelligence; but, in both cases, the manifestations of mental activity are, from a physiological point of view, the products of like nervous processes and molecular changes. If the operations of mind in man appear to us so variable as to be incalculable, and to render it often quite impossible to predict what they may be in any particular case, this uncertainty is due to our ignorance of all the factors and countless impulses which combine to produce them. In this respect, a mental resultant does not differ essentially from a mechanical resultant, and would be found on analysis to be the exact equivalent of all the motive energies which enter into its composition. But these energies are so manifold in their complexity and so mysterious in their workings that it would be impossible for any intelligence, not endowed with omniscience, to detect and determine them. The fact that mental actions are unforeseeable is therefore no proof that they are not fixed and inevitable. Man is a free agent when he acts without constraint upon the exercise of his will; but there is no such thing as free agency, if this term is used as referring to the origination of the will itself. The individual is conscious of acting according to his wishes; but he is not, and never can be, fully conscious of the forces which cause him to wish one thing rather than another, since these are often prenatal proclivities and idiosyncrasies, hereditary peculiarities of temperament running in the blood and remote from the domain of consciousness, or inbred predispositions, which in many cases he can neither know nor resist.

An appeal to consciousness as a means of explaining the real nature of psychical phenomena is as superficial and fallacious as an appeal to the senses as a means of explaining the real nature of physical phenomena. Whether applied to the microcosm or to the macrocosm, the method is the same, and the inferences are in both cases equally unsafe and delusive. The psychologist who asserts that he is free, because he feels himself to be so, is, in his logical processes of thought, a survival of the physicist who maintained that the earth is a flat and stationary body round which the sun revolves, because he saw it to be so. To accept such evidence as final and irreversible is as fatal to the progress of psychology to-day as it was for many centuries to the progress of astronomy.

It is foreign to my present purpose to discuss the question of human freedom or human necessity. I simply desire to show that whatever considerations may be adduced in favour of either hypothesis apply alike to man and to the lower animals. If Descartes declared brutes to be mere machines, La Mettrie had no difficulty, by following the same line of reasoning, in pushing his argument to its legitimate conclusion, and proving the same to be as true of human beings.

In plants, too, we not only detect rudiments of consciousness and indications of something like volition, but also discover traces of nervous organization manifesting itself in sensitiveness to irritation. *Infusoria,* polyps, sea-anemones, holothures, and other radiates distinguish between edible and inedible, or palatable or unpalatable objects, in their selection of food. Their power of choice, so far as it goes, does not differ in the manner of its exercise from that of the most fastidious gourmand. The sea pudding is, in this respect, the peer of the daintiest diner-out that ever stretched his elegant legs under the mahogany.

The eminent zoöpsychologist, Wilhelm Wundt, affirms, as one of the points which modern science has settled beyond a peradventure, the fact that the faculty of perception in the lower animals differs from that of man only in degree. He discovers between man and brute no broader and deeper chasm than between brutes themselves. All animated organisms form a chain of homogeneous beings, which are firmly linked together, and in which there is no break. Even the immense intellectual changes which man has undergone, corresponding to the growth of his brain in size and structural complexity, are the results of gradual development, and not due, in any sense, to a new departure. An obsolete psychology, with its arbitrary divisions of the mental faculties into many categories, has always been fond of drawing fanciful lines of demarcation between them; but now that we have come to recognise all spiritual life as a continuous whole, we must accept every living thing as a constituent part of this great whole. Drawing conclusions and forming judgments are elementary psychical processes, and belong to the very earliest stages of conscious life as the factors of the highest intellectual powers.

Descending still lower in the scale of animate and organic existence, we find that the closest microscopic observation, with the help of the most powerful magnifying lenses, has not yet enabled the naturalist to establish a clear and precise boundary line between the animal and vegetable kingdoms, or to set up a criterion for determining with any degree of certainty what organisms belong to each. This difficulty has led to the recognition of a third group of organisms, or vital substances, called *protista*, which are neither animals nor plants, but form, as it were, the homogeneous and protoplasmic material out of which both are evolved.

But even here the lines of separation between *protista* and plants, on the one hand, and animals, on the other hand, are by no means distinct and well defined, proving how gradually and imperceptibly the realms of Nature, as we call them, all merge into each other, and have really no existence except as modes of thinking in the mind of man.

These most primogenial of all creatures, the *protista*, although apparently mere clots of albumen without organs of sense or any sort of nervous system, are not only perceptibly affected by light, but are also attracted by different substances, selecting those which they prefer for nutriment, and showing remarkable activity and even considerable energy and ingenuity in procuring their food.

There are organisms which begin their life as plants and finally develop into animals; and there are others which undergo a reverse transformation from animals into plants, being at first endowed with locomotion, and afterward becoming stationary and taking root. *Infusoria* are thus metamorphosed into *algæ*.

How a structureless mass of matter becomes endowed with sensation and the power of propagation, and is thus changed from a chemical compound into a living creature, is a mystery which neither the dogma of divine creation nor the doctrine of spontaneous generation suffices to clear up and make perfectly comprehensible. Only analogy can throw any light upon the genesis and evolution of organic life. We know that environing influences induce inorganic or amorphous substances to crystallize: why may not favouring influences also vitalize them? We observe that changes of environment cause many species of animals and plants to thrive, to decline, and even to become extinct: why may not environment, heat, light, moisture, and other propitious conditions have originated the first germs of life? If living beings were produced arbitrarily by a creative fiat, there is no reason why they should ever undergo transformations of any kind in consequence of changes in their external conditions, or should ever die out except in obedience to a destructive fiat. The fact that they do suffer variations and become extinct and are superseded by other organisms, as the result of a change of environment, would naturally suggest that they began their existence as products of environment; in other words, that they spontaneously appeared when the proper originary conditions were realized. It is also in accordance with the theory of the spontaneous generation of organic life that there should be no break in the continuity of its development from the lowest to the highest forms.

Given a piece of protoplasm, and science is competent to derive from it all living organisms from the monad to man. The problem now presented to the biologist for solution is the production of protoplasm, or the discovery of the circumstances and conditions under which a chemical compound becomes sensitive and reproductive.

We have already seen that amorphous matter, when acted upon by certain forces—such as light, heat, cold, or sudden movement—becomes crystalline, and that these crystals have the power of reproducing themselves. Water, if perfectly at rest, may be reduced to a temperature below the freezing-point, and still remain fluid; but the slightest jar will crystallize it into ice. This phenomenon of crystallogenic attraction is a mystery, but, nevertheless, a well-recognised fact. Again, if a crystal is brought into contact with amorphous matter under proper conditions, it propagates itself, converting the mass into crystals after its kind. The physicist understands the nature and process of crystallization as little as the biologist does the nature and process of primitive germination. If the conditions are present in the one case, the crystal appears; and, if the conditions are present in the other case, the germ appears. This is all that can be said about it. But the inexplicability of either process can not be urged as an argument against its actuality.

That a plant or an animal may assimilate elements from the water, earth, and air, and use them to build up its own peculiar cell-structure, is neither more nor less intelligible than that a crystal, when placed in a proper solution, should change it into crystalline structure similar to its own. The gradual development of a living organism out of undifferentiated *plasma* in response to appropriate stimuli, such as heat, light, moisture, and electrical energy, is as easily conceivable as that two gases, hydrogen and oxygen, should combine in the form of water, which again, under the action of heat, vaporizes and disappears as steam.

In 1827 the Swedish chemist Berzelius declared that " we shall never be able to make in the laboratory any of the products of vital forces." Shortly afterward his pupil, Wöhler, disproved this so positive assertion by the chemical production of *urea;* and since that time quite a number of the products of vital force, such as indigo, salicine, and alizarine, have been fabricated chemically, and sometimes in so great quantities and with so little expense as almost wholly to supersede the natural products as articles of commerce. The synthetic chemist can even produce some of the crystals (quartz, rubies, spinels, and simili distinguishable from real diamonds only by experts) which in Nature's laboratory it took ages to form and to endow with their peculiar structure and marvellous beauty. These facts show the progress which science has made during the last half century in discovering the secrets of Nature and in imitating her mysterious processes; and there is no apparent reason why the creation of the products of vital force should not be followed by the production of vital force itself, and the artificial genesis of the germs of life.

Not only are the physical antecedents of psychical phenomena, but also the impulses and adjustive movements resulting in mental activity, the same in the lower animals and in man. Perhaps the most comprehensive classification of these impulses is that given by Dr. G. H. Schneider (Der thierische Wille, Leipzig, 1880), who distributes them into four categories: impulses of sensation (*Empfindungstriebe*), impulses of perception (*Wahrnehmungstriebe*), impulses of conception (*Vorstellungstriebe*), and impulses of thought (*Gedankentriebe*), or ideation.

Back of them all, however, lies the great original source and efficient cause of organic activity and intellectual life in its multiform manifestations; namely, the nutritive impulse (*Ernährungstrieb*), or the craving for food. Every expression of feeling, every exercise of the will, every exhibition of intelligence in the lower animals and in man, can be traced to hunger as its fountain-head. From the pressure of hunger and the desire to prevent its recurrence spring the love of acquisition, the systematic accumulation of wealth, the idea of ownership in things, or the general conception of personal property, which is the strongest cement of social and domestic life, codes of laws and systems of morals, discoveries, inventions, industrial and commercial enterprises, scientific researches, and the highest achievements of culture and civilization.

It is true that, as a man rises in the scale of intelligence, other and nobler incentives to activity come into operation and act even more powerfully than the primal nutritive impulse. The latter, however, always asserts and insists upon the priority of its claims; and not until these have been satisfied and the stress of hunger relieved, and in some degree permanently guarded against, does the individual think of devoting his energies to higher pursuits. Spinoza had to secure his subsistence by grinding his stent of lenses before he could gratify his love of philosophy and find leisure to work out the ethical and metaphysical problems in the solution of which all his intellectual powers were engaged. It was the chief grievance of Xantippe that her husband would waste his time in getting up, according to what has since been known as the Socratic method, unprofitable "corners" in speculative questions, which brought in no pecuniary returns, and neither kept the pot boiling nor contributed to the alimentary worth of its contents. Still, it is highly probable that the nutritive impulse would have been stronger in the Grecian sage if he had been thrown upon his own resources for subsistence, and had not relied upon the sufficient persistency of this natural instinct in his spirited spouse to supply the wants of a modest Athenian household.

Parallels to this feature of the conjugal life of Socrates are found in many a New England village of to-day, where we see the exuberant practical energy of the wife repressing the easy-going, wool-gathering husband, and reducing him first to a domestic nullity, and finally to a confirmed loafer and peripatetic philosopher, sententious and seedy, wise and worthless, loved and laughed at by all men.

The final purpose of the nutritive impulse and of the various subsidiary impulses which minister to it is the preservation of the species. Seeking food, fighting foes, forming friendships, sexual attraction, care of offspring, social feeling, love, hatred, fear, jealousy, cruelty, kindness, revenge, deceit, "all thoughts, all passions, all delights," are subservient to this one great end.

Not only is the preservation of the species the aim of all the energies developed by animal organisms in their present state of being, but it is also the genesis of the belief in a life to come. The doctrine of the immortality of the soul springs from man's unwillingness to give up the struggle for existence, even after the dissolution of his physical frame. It is the expression of his antipathy to annihilation and his longing to live and to develop to a still higher degree his spiritual powers. The soul is the ideal of individuality in its purest form, just as the gods of a people are its ideals of humanity in its purest form, although it may be, as Dr. Svoboda remarks, that "a soul which no one remembers is as devoid of reality as a god which no one worships."

Impulses of sensation are produced by immediate contact of the living organism with external objects;

impulses of perception are called forth by seeing objects at a greater or less distance; impulses of conception originate in the presentation of real but absent objects to the mind by the power of memory; impulses of thought may arise out of the mere imagination of objects or the simple apprehension of things not actually existing. There is, however, no break in this series of cognitive movements, from the most automatic reflex action to the most complex processes of abstraction and generalization; nor is it possible to determine how far they are due to mental and to non-mental factors, or to draw a boundary line defining the limits of each. We know that thought and emotion are always connected with certain molecular movements in the brain. Whether the cerebral movements are the cause or merely the concomitants of the mental manifestations we can not tell. All that we can assert is that, within the limits of our experience, the latter are inseparable from the former, and wholly dependent upon them.

In the lower animals the lower impulses are predominant, and this predominance is used by Schneider, in a general way, as the basis of psychological classification. Thus he regards *protozoa* and radiates as sensation animals; the mollusks and articulates as perception animals; the vertebrates, with exception of the human species, as conception animals; and man as pre-eminently a thought animal. Nevertheless, it is to be observed that man acts in obedience to all these impulses, and that the lower animals, which are usually governed by impulses of sensation, perception, or conception, may and do exercise thought, and are influenced by imagination and reason.

Oken regarded the life of the lower animals as a sort

of mesmeric state, due to the ascendency of the sleeping soul located in the liver over the waking soul with its seat in the brain, and classified them according to their supposed temperaments into melancholy, sanguine, and choleric. The first class comprises fishes and reptiles; the second, birds; and the third, mammals. Animals of the first class have memory and sensation only; those of the second class have perception, conception, and concrete ideas; those of the third class have understanding, intelligence, and reason, but not self-consciousness, which is the sole attribute of man. This classification, although superficially suggestive of that proposed by Schneider, is loose and unscientific; and, instead of being based upon accurate observation, it is made to suit certain mystical notions and metaphysical theories.

Children, savages, and the rude and ignorant classes of civilized society yield more readily than highly developed races and individuals to the lower impulses of sensation and perception, as is evident from their lack of self-restraint when excited by the presence of desirable objects, and their disposition to gratify their appetites without thought of the future.

Schneider maintains that squirrels, hamsters, and woodchucks, in collecting and storing food, act solely in obedience to the impulses of perception and conception. Thus the perception of a nut causes them to pick it up; the conception of their hole or burrow impels them to carry off the nut; and, when they have reached their abode, the perception of the place induces them to lay it down or store it. Such an explanation, however, really explains nothing, and is wholly inadequate to account for the animal's conduct. It does not furnish any sufficient motive for the action, and therefore leaves it as unintelligible as it was before. If these creatures exercise no foresight, and have no notion that the nut is to serve them as food in the coming winter, they would, if hungry, eat it at once; if not hungry, they would let it alone; they surely would not store it for future use. No squirrel is tempted by a fair exterior to try his teeth on a hollow nut, or to add worthless material of this sort to his winter supply of provisions. But, if he were governed solely by the aforesaid impulses, he would not be capable of such discrimination. The mental process which leads him to discard every nut that has not a sound kernel in it is not confined to a simple impulse of perception. The fact, too, that he does not merely pick up the nuts which he happens to find in his wanderings, but sets out in search of them, proves that the action is due to the exercise of thought, and that the agent is clearly conscious of the purpose for which it is performed.

The propensity of the carrion fly, on the other hand, to lay its eggs in putrefying flesh, which will supply its young with proper nourishment, is stimulated and directed wholly by the impulse of perception, and especially by the sense of smell, since it often lays its eggs on plants which have the odour of carrion, but not its nutritive qualities, so that the young perish from lack of food as soon as they are hatched. Again, the tumblebug, on perceiving a small, round object, is seized with an irresistible impulse to roll it, although it may be a piece of wood or stone instead of a ball of dung containing its eggs. But this sort of fatuity is, unfortunately, not confined to tumblebugs. One meets with many persons in daily life who are easily deceived by outward semblances, think they know a round and rollable thing when they see it, and are constantly engaging in all sorts of foolish and unfruitful enterprises. In forming ties of friendship, love, and matrimony, and in entering into a great variety of social relations, young people are especially apt to be led by mere impulses of perception and conception, so-called fancies, which are often blind and irrational whimseys of the most delusive and pernicious character.

When a fox sees the bait of a trap, there are two distinct impulses immediately excited in Reynard's breast—the impulse due to perception, which tempts him to seize the tempting morsel, and the impulse due to conception, which suggests the danger of being caught; and his safety depends upon the comparative strength of these two impulses. Under such circumstances, a young animal will probably yield to the perception impulse, and fall into the snare: whereas an older and wilier fox will most likely be governed by the conception impulse, whereby the sense of peril overrules the strength of appetite, and will thus escape.

Feigning death in the presence of danger implies not only a clear conception of the impending peril, but also a remarkable degree of cunning and self-control in evading it. The theory of Prof. Preyer that this simulation is simply a state of catalepsy produced by the paralyzing effects of fear is wholly inadmissible, and will never be accepted by any one who has seen an opossum "playing 'possum" or read Audubon's vivid description of such a performance. No disciple of George Fox ever developed the power of passive resistance possessed by the opossum. The female *didelphys* is a heroic mother, and will calmly suffer martyrdom for the safety of her offspring. She can open at will the pouch in which she keeps her young, but no amount of torture can force her to do so. To get them out is as difficult as to get a joke into a Scotchman's head, and can be effected only by the same means—a surgical operation. Hypocrisy (that is, "acting") is a trait shown by all weak animals in self-defence. Dogs are adepts in putting on an air of innocence when they are fully sensible of having done wrong, and in craving pardon by expressions of mingled contrition and flattery when their guilt has been detected and exposed.

Birds and mammals, which live in flocks and herds, post sentinels, when they are feeding or sleeping or engaged in any perilous enterprise, in order to warn the community of the approach of an enemy. Flamingoes, wild geese, turkeys, gulls, bustards, crows, ravens, storks, prairie hens and prairie dogs, monkeys, zebras, wild horses, chamois, beavers, otters, walruses—in short, all gregarious animals—have this habit. The sentinels also show great discrimination in the discharge of their duty, paying no heed to harmless animals like a sheep or a cow, but sounding an alarm at the approach of a beast of prey or a man. Before migrating to any particular place, spies are sent out to ascertain whether the change would be desirable or attended with danger. In Siberia deputations of squirrels go on such missions, usually in August, crossing dreary wastes, swimming rivers, and enduring all sorts of hardships until they reach the high plateaus of the pine forests. In a few weeks they return, report on the prospects of the cone harvest, and toward the end of September guide the whole squirrel community to the most favourable spot. The zeal of these emissaries in the performance of their task is shown by the bruised and blistered condition of their feet; and they appreciate the importance of their office as fully as did the men whom Moses sent to spy out the land of Canaan.

It is also a curious fact that snipes, stilts, and other birds which frequent the river banks and the seashore do not keep sentry themselves, but rely for security on the vigilance of the plover, which is quick to signal any danger. For the same reason zebras are fond of feeding near ostriches, where they are free from all anxiety, knowing that the ostriches are always on the alert and quick to scent the slightest suspicion of an approaching foe.

That these actions are performed with a full consciousness of the object to be attained is undeniable, and can be explained on no other theory. Stationing sentinels indicates not only a high degree of foresight and forethought, but also gives evidence of remarkable moral qualities, as the expression of individual self-sacrifice for the common good, or what in human societies would be called public spirit or patriotic sentiment. Sentinels and spies expose their lives for the safety of the flock or herd. It makes no difference whether they undertake this service voluntarily or are compelled to perform it: the existence of such an office marks a high development of the moral sense in the perception of the obligations of the individual to the community of which he is a member.

Even cold-blooded sea creatures know how to profit by associations of this kind. Little sea crabs seek protection in the vicinity of the polyp from their arch-enemy, the squid. In like manner the pilot fish is safe from the attacks of the tunny in the neighbourhood of the shark. Here it is self-preservation, and not friendship, that forms the bond of association. It is the same with finches and sparrows, which take refuge from falcons in the eyries of eagles. In such cases, the weaker animal is protected by the mere presence of the stronger, but there is no evidence that the latter derives any advantage from the companionship.

Usually, however, the relation is one of mutual benefit, as, for example, in what might be called the love of the hermit crab for the sea anemone. The hermit crab takes up its abode in the abandoned shell of a mollusk to which a sea anemone is attached. If it wishes to change its habitation, it takes the sea anemone with it; or, if it finds a suitable shell without a sea anemone, it goes in search of this companion, who both adorns and protects its home—adorning it like a flower of rosy hue, and protecting it with its mesenteric filaments that sting whatever they touch, and thus ward off the assaults of fish which would otherwise drag the hermit crab from its shell and devour it. In return for this kindness the hermit crab provides the sea anemone with food. In this union each seeks its self-interest and secures its highest good.

A queer kind of fish is the stargazer, or uranoscope, so called because its eyes are on the top of its head and are therefore always looking heavenward. On account of this sanctimonious look it is also known as the "sea parson." This fish, which Hippocrates prized as wholesome food, probably because its flesh is especially offensive, and Konrad Gessner more than two centuries ago characterized as "very dreary and dreadful to look upon," lives for the most part buried in the mire or sand, with nothing visible but its staring eyes and vertical mouth, from which projects a long, cylindrical, cartilaginous flap that wriggles like a worm. No sooner does one of the fry of little fish that gather round this supposed worm bite at it than it is seized by the treacherous flap, and disappears in the pitlike mouth of the hidden stargazer.

Now, as to the mental process here involved, Schneider maintains that the stargazer buries itself in the slime in obedience to an impulse of perception, stretches out its squirming flap in obedience to an impulse of conception (i. e., of its prey), and draws it in again with the captured minnow in obedience to an impulse of sensation, but that it has no consciousness of the purpose for which it performs all these actions. The naturalist deems himself justified in this summary treatment of the psychology of the subject, simply because he is dealing with a creature of low organization, and is unwilling to admit that its thoughts can be as his thoughts, even when there is a striking resemblance in their external acts. The most accomplished angler that ever whipped a stream obeys a mere impulse of sensation when he hooks his fish, although he may exercise his reason in resisting this impulse, and not respond to every nibble at the bait as an inexperienced fisher would do. For aught we know, the stargazer may use the same discretion. Reptiles, birds, and mammals of many kinds, toads, scorpions, crocodiles, herons, crakes, dogs, cats, lions, tigers, and human beings lie in wait for their prey. The man is well aware of the purpose for which he lurks in ambush, and the same is true of the tiger and the cat. Indeed, all the way down in the scale of predatory animals, from the savage to the sea devil and the stargazer, there is no point at which the action ceases to be conscious and rational, and becomes purely sensational and automatic. In the higher organisms the higher faculties predominate, and in the lower organisms the lower faculties; but in all of them, from the highest to the lowest, the action is the resultant of impulses of sensation, perception, conception, and thought variously combined and inextricably blended.

A typical illustration of the illogical inferences drawn by psychologists as to the mental powers of the lower animals is furnished by an interesting experiment made by Mr. Amtsberg, of Stralsund, and reported by Dr. Möbius to the Society of Natural Science for Schleswig-Holstein, in 1873. A large pike, which was wont to devour the small fish in an aquarium, was finally separated from them by a plate of glass, so that, whenever he attempted to seize his prey, he struck his snout so violently against the transparent barrier as to be quite stunned by the blow. Nevertheless, he kept up these attacks for some time. At length, however, they became rarer, and finally, after three months of disheartening effort, ceased altogether. After the lapse of six months the glass plate was removed, and the pike swam about freely among the other fish without attempting to eat them. But no sooner was a strange fish put into the aquarium than he gobbled it up.

In the opinion of Dr. Möbius and other scholars who have accepted his interpretation of the phenomena, the conduct of the pike " was not based on judgment," but was the result merely of " the establishment of a certain direction of the will in consequence of a series of uniformly recurring sensuous impressions." But this holds true of all discipline, and is precisely the process by which the judgments of children, and indeed of the great majority of adults, are formed. There are hosts of persons who go through life constantly bumping their heads against invisible walls and learning wisdom—if they learn it at all—only by hard knocks. Very many lack even the perception shown by the pike, and do not know when a spiritual barrier has been taken away and the sphere of their intellectual activity enlarged, but continue to move along the line of the old partition wall, and never dare to go beyond it.

In the case of the pike, the glass plate was simply the means of inculcating a definite idea; namely, that certain fish were not to be eaten. Every blow against the unseen barrier was an admonition and injunction on this point, and a vigorous enforcement of the lesson to be taught, just as a wilful child learns to let forbidden things alone by a smart slap on the fingers. To affirm that "the pike acted without reflection," or that it was "a machine with a soul, which has this advantage over soulless machines, that it can adapt itself to unforeseen circumstances," and that "the plate of glass was to the organism of the pike one of these unforeseen circumstances," is to make a terrible pother of words without sense, and to give an explanation that explains nothing. A machine with a soul is a contradiction in terms, since an organism with a soul ceases by virtue of this endowment to be a machine.

Nor was it a "mark of stupidity" that the pike did not eat the fish after the plate of glass had been removed, but rather an indication of docility and discrimination. The ability to distinguish between the fish that were not to be eaten and those that might be eaten shows close observation and the power of comparing objects and discerning their relations and properties; and what is this but judgment? A certain association of ideas was established in the pike's mind by the intervening plate of glass, just as it may be established in a child's mind by an intervening slap. The law of mental action is in both cases identical. In the instance adduced, it required months of discipline to establish this association of ideas, but it was so firmly fixed as never afterward to be broken. The lesson once learned was not forgotten, and in this particular the pike's education was complete. He was trained up in the way he should go, and did not depart from it. It is precisely in this manner—namely, by threats and blows—that a cat is taught not to touch caged birds. The naturalist Lenz tells of an old tabby which, having been thus trained, imparted the instruction to her kittens in the some way, cuffing them whenever they approached the cage with the feline stealth indicative of felonious intent.

By an application of the same principle, a horse is taught to stop at the word "whoa"; namely, by attaching a rein to the animal's foot and pulling the foot clear of the ground every time the word "whoa" is uttered. The horse is thereby forced to stop, and thus learns what "whoa" means, and acts accordingly. It is not necessary to suppose that such an absolute connection is established between the sound "whoa" and the pulling of the forefoot from the ground as to make the horse think of the act whenever he hears the word. If this were the case, the animal, on hearing the word, would not only stop, but also lift the forefoot. The procedure is purely didactic. The horse learns what his master means by "whoa," and obeys, but no longer thinks of how he came to learn the lesson, any more than a man, who in his youth was compelled to study Latin by the application of the rod, thinks of "the threatening twigs of birch" whenever he reads a Horatian ode or a Virgilian eclogue.

Few persons have any conception of the pains taken by a parent bird to teach her little ones how to get their living and to make their way in the world. Thus, for example, a mother sparrow may be often seen, at the proper season seated with her fledgelings on the ridge of a roof and letting a pea, berry, or round piece of bread roll down into the eaves-trough. She repeats this performance until one of the most precocious and alert of the brood takes part in the game, and very soon the whole family join in the sport, hopping after the rolling object and vying with each other in securing it. The mother now varies the performance by catching the thing before it reaches the gutter. After a time the young birds succeed in this more difficult exploit, and thus take their first lesson in the art of seizing moving objects. C'est le premier pas qui coûte. The skill acquired in capturing a rolling bread-crumb is easily applied to a flying bug.

The way in which many psychologists talk about the mental faculties of animals recalls Heine's interview with the old lizard at Lucca. In the discussion which ensued, the poet dropped the words "I think." "Think!" cried the lizard, with a sharp, aristocratic tone of profound contempt; "think! which of you thinks? For three thousand years, wise sir, I have investigated the spiritual functions of animals, and have made men and apes the special objects of my study.

I have devoted myself to these queer creatures with as great zeal and diligence as Lyonnet to his caterpillars; and, as the result of my researches, I can assure you that no man thinks. Now and then something occurs to him; and these accidentally occurring somethings he calls thoughts, and stringing them together he calls thinking. But you can take my word for it, no man thinks; no philosopher thinks; neither Schelling nor Hegel ever thought; and, so far as their philosophy is concerned, it is mere air and water, like vapours in the sky. I have already seen countless successions of these clouds floating proudly and securely over my head, and the next morning's sun dissolved them into their original nothingness. There is, in reality, but one true philosophy, and that is engraven in eternal hieroglyphics on my own tail." This lordly and disdainful attitude of the venerable saurian toward the human race is a witty persiflage of the anthropocentric conceit which perverts man's views of his relations to the lower animals.

If a writer, with the critical acumen of Gervinus, asserts that the nations of antiquity "took no delight in Nature," and Schiller affirms that "Nature interested the understanding and excited the curiosity of the Greeks, but did not awaken in them any moral feeling," if keen thinkers thus fail to get a clear and correct appreciation of the mental and emotional capacities of their fellow-men in earlier epochs and more primitive stages of intellectual development, how much more difficult must it be to analyze and estimate aright the psychical phenomena of animal life that lie still remoter from our own!

It is a significant circumstance that metaphysicians

have never made any valuable contributions to zoöpsychology. This is because they have always discussed the mental constitution of animals without having adequately observed their habits, or have endeavoured to make such facts as came within their range of observation fit into some preconceived theory, discarding as worthless whatever could find no place in the systems of thought they were pledged to uphold. Wild speculation on a small amount of real capital is apt to prove as disastrous in the province of philosophy as on the stock exchange.

Aristotle, who was perhaps less liable to this reproach and dealt more with positives than any of his contemporaries, maintained, nevertheless, that heart-beating is a phenomenon peculiar to man, "because he alone is moved by hope and expectation." The process of reasoning which led the Stagirite to this absurd conclusion seems to have been something as follows: Fearful or pleasurable anticipation causes the heart to beat, and this pulsation can occur only where such feelings exist. The lower animals, however, live wholly in the present, do not look forward to the future, and are not agitated by pleasant or painful presentiments: *therefore*, their hearts do not beat. Such is the vicious circle in which the greatest logician and clearest thinker of antiquity allowed himself to be caught, through excessive confidence in the validity of syllogisms and the lack of a little observation. An Athenian boy with a bird in his hand would have put him to shame, and laughed his logic to scorn. Cassiodorus held that the life of the lower animals resides in the blood, whereas the *anima*, or soul, is a principle peculiar to man, and distinct from the blood, and based this fanciful theory on a false etymology: *anima quasi ἄναιμα, id est a sanguine longe discreta,* which would identify spirituality with *anæmia.*

Lactantius, in discussing the origin of error (Inst. Div. Lib. II. cap 10), makes a still more subtile and strained distinction between men and brutes. "For we," he says, "being a heavenly and immortal race, make use of fire, which is given to us as a proof of immortality, since fire is from heaven; and its nature, inasmuch as it rises upward, contains the principle of life. But the lower animals, inasmuch as they are altogether mortal, make use of water only, which is a corporeal and earthly element, and because of its unstable nature and downward tendency, shows a figure of death. Therefore, the cattle do not look up to heaven, nor do they entertain religious sentiments, since the use of fire is removed from them." Elsewhere in the same apology (II. 1) he states as a significant fact that the Greeks called man ἄνθρωπος because he looks upward. It is strange how much stress has been laid upon this false etymology (for the word means man-faced, and contains no suggestion of looking upward), what far-reaching physiological inferences have been drawn from it, and for how many centuries poets have not ceased to ring changes upon it. Looking upward is, as we have already seen, a physiological peculiarity of the stargazer and the sea devil, but not of man, who naturally looks straightforward, and can look upward, as Galen remarked more than sixteen centuries ago, only by painfully bending back his head. The goose is infinitely his superior in the ease with which it can turn its eyes heavenward.

Yet Ovid says—

> Pronaque cum spectent animalia cetera terram,
> Os homini sublime dedit, cœlumque tueri
> Jussit; et erectos ad sidera tollere vultus.

Less than a century later Silius Italicus, in his epic of the Second Punic War (xv.), amplified the verses of his precursor and prototype as follows:

> Nonne vides hominum ut celsos ad sidera vultus,
> Sustulerit Deus, ac sublimia finxerit ora?
> Cum pecudes, volucrumque genus, formasque ferarum,
> Segnem atque obscenam passim stravisset in alvum.

Racine repeats the same thought in the lines—

> L'homme élève un front noble et regarde les cieux,

and Milton embodies it in his description of creation in still fuller and more poetic form:

> There wanted yet the master-work, the end
> Of all yet done; a creature who, not prone
> And brute as other creatures, but endued
> With sanctity of Reason, might erect
> His stature, and upright with front serene
> Govern the rest, self-knowing; and from thence,
> Magnanimous, to correspond with heaven.

Cowper sings the same strain:

> Brutes graze the mountain-top with faces prone
> And eyes intent upon the scanty herb
> It yields them,

as though the hungry savage were any less "intent" upon the food with which he gluts his maw.

Birds not only stand erect, but also, by the power of flight, free themselves more than mammals from bondage to the earth. "But," says Steinthal, "however high they may soar, they still belong to the earth." The same is true of man, however far

> His piercing eyes erect appear to view
> Superior worlds, and look all Nature through.

The ant is not inferior to the bee in intelligence because it crawls on the ground instead of hovering in the air. The owl surpasses man in the facility and freedom with which it can turn its head in every direction, but this flexibility of its neck does not contribute in any degree to the enlargement of its mental horizon. Steinthal compares the brute to a piece of cloth fastened at all four corners to the ground, whereas man is like a piece of cloth attached at only two points, so that the greater part of it flutters in the air. "The influence of this power of free motion," he adds, "in promoting the development of intellectuality is incalculable." This rather vulgar comparison of man to a flapping sheet, recalling somewhat ludicrously the possibility of his being "two sheets in the wind," does not illustrate in the least the point in question.

Man's superiority of bodily structure and constitution in respect to his mental development was very succinctly stated by Herbart nearly a century ago, as follows: "He has hands; he has speech; he lives through a long, helpless childhood." (*Er hat Hände; er hat Sprache; er durchlebt eine lange, hülflose Kindheit.* Werke, vi. p. 206.) The last-mentioned point has been taken up and most fully and satisfactorily worked out by Mr. John Fiske. Animals without hands, or prehensile organs that may by use be converted into hands, derive no intellectual advantage whatever from an upright position. The penguin may have the habit of standing erect on its feet and flopping its quill-less wings for countless generations without adding in the least to the size or complexity of its brain. The assumption and permanent maintenance of an upright posture marked an epoch in the evolution of the human race, only as contributing to the differentiation of the hand from the foot, and to the development of the former as an organ of investigation instead of a means of locomotion.

The form and flexibility of the hand and the extreme delicacy of the sense of touch, especially in the tips of the fingers, are the chief sources of man's intellectual progress, so far as this is dependent upon his physical structure. The capability of grasping an object with firmness and precision and holding it with ease and exactness in a variety of positions not only renders possible the use of tools, the acquisition of mechanical skill, and the growth of the arts, but also exerts a direct influence upon the intellect by cultivating the powers of close observation and intense concentration of thought. It is by no means a mere accidental coincidence that many words used to denote operations of the mind are spiritualizations of the functions of the hand; as, for example, when we speak of grasping or handling a subject, seizing a point, catching an idea, and comprehending a proposition. These expressions, now employed as simple figures of speech, are records of real facts and natural processes in the early educational history of mankind, since it was by the frequent repetition of the manual action that the higher and fuller mental life of the individual was developed and the progress of the race promoted.

The mental and moral value of mechanical labour as a discipline for the young is now just beginning to be appreciated and to be assigned its proper place in pedagogics. The boy who has learned to draw a straight line has learned a lesson in rectitude; and in making a box or a table he builds up his own character, and gives it additional symmetry and stability. The influences which civilized the race in its infancy are still the most efficient agencies in civilizing each individual; for, notwithstanding the hereditary transmission of culture, as yet every healthy child is born into the world more or less a savage.

CHAPTER VI.

PROGRESS AND PERFECTIBILITY IN THE LOWER ANIMALS.

Animal and human institutions. Domestic and social life of beasts and birds. Bee colonies. Improvements in nest building. Architectural skill of ants and termites. Destructive energy of the latter. Immense size of their mounds. Artificial comb foundation for bees. Perfectibility of the species. Effects of specialization in training. Influence of domestication. Schutz's theory of animals as puppets of higher powers.

WHAT we call institutions are only organized and hereditary instincts, and are common to man and the lower animals. The original social character of animals, which forms the basis of their institutions, is also the quality that renders them capable of domestication. Man simply takes advantage of this quality, and turns it to his own account by bringing the animal into his own domestic circle and service and making it a member of his household.

In birds, for example, the conjugal instinct is remarkably strong, or, as we would say in speaking of human relations, the institution of marriage, either in its monogamous or polygamous form, is firmly established and highly developed, and forms the foundation of a well-ordered domestic and social life.

The paternal fox trains his young with as much care and conscientiousness as any human father; the beaver constructs his habitation with the foresight of a military engineer and the skill of an experienced architect; the bee lives in well-regulated communities, forms states, and founds colonies; and the ant not only cultivates the soil, plants crops, gathers in the fruits of his labour and stores them for future use, and keeps other insects as domestic cattle, but shares also the vicious propensities and domineering disposition of man, waging war on creatures of his own species and holding his prisoners as slaves.

These habits or customs have the same origin and character in the lower animals as in man, being in both cases products of evolution and undergoing modifications from generation to generation. Animal, not less than human, societies are governed by their laws and traditions, and preserve a sort of historical continuity by which past and present are bound together in a certain orderly sequence. Beehives which suffer from over-population rear a new queen and send forth with the old one a swarm of emigrants to colonize, and the relations of the mother-hive to her colonies are known to be much closer and more cordial than those which she sustains to apian communities with which she has no genetic connection. Here the ties of kinship are as strong and clearly recognised as they are between consanguineous tribes of men.

Again, the statement that animal habits are fixed, and human customs variable and improvable, is true only to a very limited extent. Closer observation has shown the latter to be more stable and the former more mutable than is generally imagined, especially if we compare the highest orders of animals with the lowest human tribes. In primitive society and among savage races customs remain the same for countless generations, and seem to be quite as persistent and incapable of change as animal instincts.

Not only do animals, often in the course of a comparatively short period, undergo marvellous transformations both of mind and body, through the force of natural selection or by careful interbreeding, but they are also led by circumstances and through forethought to make conscious and intentional changes in their manner of life.

It is curious to note the variety of characteristics distinguishing members of the same family or genus. Thus, the European cuckoo lays her eggs in the nests of other birds, and leads the life of a shiftless parasite and shameless polyandrous vagabond. The American cuckoo, on the contrary, has not yet learned to shirk her maternal duties and domestic responsibilities, but, like an honest and thrifty housewife and conscientious mother, hatches her own eggs and rears her own young. The South African and Australasian representatives of the *cuculinæ* follow, in this respect, the habits of the European bird. There is also a species of *molothrus*, which sometimes begins but seldom finishes a nest, like the hypothetical man in the parable, who would fain build without first sitting down to count the cost. She is seized occasionally with a spasm of virtuous endeavour in this direction, but soon yields to the greater comfort and convenience of imposing upon others the burden of brooding and nurturing her offspring. Evidently she turns the matter over in her mind, and, like Rousseau, reasons herself into the belief that it is better not to assume any family cares, but to cast her children as foundlings upon the bosom of public charity. "There are the goldfinches, thrushes, fly-catchers, cardinal grossbeaks, and other fussy motherly fowl," she seems to say, "willing enough to undertake the charge; why not gratify their low philoprogenitive passion, and thus enable me to devote myself to more congenial pursuits!" Still another kind of *molothrus* leads the life of a squatter, never building a nest of her own, but brooding in the abandoned nest of some other bird.

Many birds have, within the memory of man, made considerable advances in architectural skill, and adopted new and improved methods of constructing their nests. This progress has been observed especially in the swallows of California since the settlement of that country, and in all cases the young profit from the knowledge acquired by their parents, and the improvement becomes a permanent possession of the race. In places where they are particularly exposed to the attacks of pugnacious sparrows, they have been known to close the opening in front of their nests and make the entrance on the back near the wall. In some instances this purely precautionary and defensive change of structure, after its efficiency had been tested in a single nest, has been adopted by the swallows of an entire district. Orioles, according to the observations of Dr. Abbott, finding that the bough from which they have suspended their nest is too slight to sustain the weight of the full brood, attach it by a long string to the branch above, fastening it securely "by a number of turns and a knot." It would be difficult to say in what respect the mental process leading to the adoption of such a mechanical contrivance differs from that which causes an architect to buttress a weak wall.

The Baltimore oriole also adapts the texture and structure of its nest to the exigencies of climate. In the Southern States it selects a site on the north side of a tree, and builds of Spanish moss loosely put together and without lining, so as to permit a free circulation of air. Farther north it seeks a sunny exposure, builds more compactly, and uses some soft material for lining. The impulse to build is instinctive, but conscious intelligence is exercised in modifying the methods of building to suit circumstances.

The same bird now uses yarn and worsted instead of vegetable fibre for its nest, but it always selects for this purpose the least conspicuous colours, such as gray and drab; and yet the bird's gorgeous plumage is proof, according to the theory of sexual attraction, that bright colours are pleasing to it. Here we have an example of æsthetic pleasure being subordinated to considerations of safety; the prudent oriole, notwithstanding its fondness for resplendent hues, choosing those colours which render its nest less visible and more difficult to discover, and rejecting those which, in other respects, are more gratifying to its fancy.

The tailor-bird of East India used to stitch the leaves of its nest together with fine grass, horsehair, and threads, which it twisted out of wool; since the introduction of British manufactures it uses sewing thread and the filaments of textile fabrics, except in remote regions, where the ingenious bird still works on in the primitive way. So, too, in America, birds in constructing their nests everywhere turn to their account the products of human industry and keep abreast with the progress of the age. The materials employed correspond to the contemporary state of civilization, and mark the periods of industrial development through which the human race has passed. The wagtails, in a watch-making district of Switzerland, have learned to build their nests of fine steel shavings; a nest of this kind, if preserved, would indicate to the inhabitants of that country a thousand years hence the kind of industry that was carried on by their ancestors. Sparrows, which usually build in chinks of walls or under roofs, if forced to build their nests in trees or other unsheltered places, cover them with a sort of hood to keep out the rain. Buffon, who records this fact, adds: *L'instinct se manifest donc ici par un sentiment presque raisonné et qui suppose au moins la comparaison de deux petites idées.* In the presence of such clear manifestations of thought and reflection, it seems absurd to speak of a "sentiment almost reasoned," or to indulge in condescending baby-talk about "two little ideas."

Apiarists now provide their hives with artificial comb foundation on which the bees build and are thus relieved of some of the labour performed by their predecessors.* Instead of gathering propolis from the buds of plants, the workers stop their hives with the mixture of resin and turpentine with which the arboriculturist salves wounded trees, and readily substitute oatmeal or the flour of wheat and rye for pollen, if they can not easily procure the latter. In countries where the flowers blossom late these surrogates are

---

* A very superior kind of comb foundation is manufactured by Gustav Ad. Friderich in Greifswald, Germany.

often provided by apiarists and placed before the hives in the early spring. In Cuba the bees pillage the sugar plantations, and in North Germany, near Stettin, the numerous sugar refineries are subject to the same depredations. In Barbadoes, where these sources of supply are accessible during the whole year, the bees gradually cease to gather honey from flowers and appropriate the products of human industry. They also visit cellars, in which kegs of syrup are stored, for the same purpose, and indulge in stolen sweets to such excess that they fall to the ground and perish, so that the apiarists suffer considerable loss. Huber gives an interesting account of the manner in which honeybees rob humblebees in times of scarcity, carrying on this spoliation systematically for several weeks until nothing is left.*

In a work entitled A Modern Bee Farm, Mr. S. Simmins describes the "sweating" methods by which practical apiarists turn the industrial virtue of bees to the best account. Large fields of white clover, borage, and sanfoin planted near the hives enable the bees to gather honey with the least possible loss of time, and it is estimated that seventy-five acres of these flowering herbs will occupy one hundred hives profitably for three months (June, July, and August), and produce ten thousand pounds of honey in a single season. The triumph of the "sweater's" art, says Mr. Simmins, is in inducing the bees to fetch this enormous quantity of honey, without neglecting the arrangements for storing it in the hives. The honey, being

---

* See Ludwig Büchner, Aus dem Geistesleben der Thiere, 4th ed. Thomas: Leipzig, 1896, pp. 322-324.

liquid, must be bottled, and the bees will only put it into comb of the exact size and texture, which instinct has taught them to make. As comb-making is much lighter and safer work than honey-gathering, with its dangers from storms, wasps, and birds, it is generally assigned to young bees, while their elders go afield. In order that as few bees as possible may remain in the hive for this purpose, the bee-keeper provides ready-made foundations for the cells, stamped in real wax and of the natural size. He also removes the combs full of honey, spins them round in a tin churn, and replaces them in the hive empty—a hint which the bees take as as invitation to refill them. The bees seem delighted to make the most of the opportunities so thoughtfully provided for them. By using the mechanically stamped "foundations" for their cells, they make a more perfect and symmetrical comb than is often constructed without help. The bottoms being regular, no "crooked comb" is ever built upon it. The size stamped is also uniformly that of worker cells; thus, there is no room for drone cells, producing bees which can not be "sweated" or made profitable in any way. Mr. Simmins thinks that the trial of this system for another twenty years "may possibly show certain strains developing a tendency to forget how to construct comb foundation, just as some breeds of fowls are forgetting how to hatch their eggs. We can not suggest an improvement in the architecture of the cells because they are mechanically perfect in economy of material and space. But the readiness with which the honeybee has accepted and incorporated in its comb the materials supplied by man would seem to indicate the possibility of further experiments to determine how far its mechanical instinct is capable of modification." It is a mistake to attribute the hexagonal structure of the cells to mechanical instinct, since it is due solely to external pressure. The melipona or native American bee constructs long and round cells which show no approach to the hexagonal form, except when they are in close contact, so that the whole is filled without interstices. Indeed this is the shape which all soft and pliable balls or cylinders take when they are pressed uniformly together. Thus, if water be poured into a bottle filled with peas, the latter, as they swell and press against each other, will gradually change from spheres into hexagons. In like manner soap bubbles produced by blowing into a basin of suds are six-sided so long as they remain in contact, but become spherical when they float off singly into the air. The originally round cells of the human body become hexagonal when pressed together in the mucous membrane, in tumours, cancerous formations, and other morbid growths. Although bees are remarkably conservative, it is evident that their methods of work may be considerably modified by human agency.

The facts already mentioned, and many others which might be adduced, suffice to prove that animals avail themselves of new discoveries and easier methods in order to increase the comforts and conveniences of life.

Haeckel asserts that the rude aboriginal ants, which lived many thousand years ago, perhaps as early as in the Chalk period, had as little idea of the advanced division of labour prevailing in the different modern ant states as our forefathers of the Stone age had of the culture of the nineteenth century. Both ants and men have worked themselves up to their present stage of development on the slow and painful path of progressive evolution. Even now there are varieties of ants which know nothing of the exact and elaborate system of division of labour found in civilized formican communities, and which bear the same relation to the latter that the rude aborigines of Africa and Australia do to the civilized nations of the present day. The same is true of the hymenoptera and arachnida, in whose habitations there is traceable a process of architectural evolution analogous to that which has taken place in the history of mankind. This is evident from a comparison of the nests of wasps and humblebees and the cells of the native American bee (*Melipona*) with the perfectly formed comb of the European honey-bee, or the habits of ordinary earth spiders with those of trapdoor spiders.

Ant hills are very complicated structures. They are partly under the earth and partly above it, and consist often of twenty to forty stories, thus relatively surpassing in size the sky-scraping edifices of modern American cities. They are constructed of pieces of wood, earth, pebbles, leaves, stems of plants, pine-needles, and other materials apparently lying promiscuously in heaps, but found on closer examination to be arranged so as to form halls, corridors, and rooms adapted to the wants of the inhabitants. It is also curious to note how they make changes in these dwellings to suit their needs. Sometimes the work of one ant will be torn down and rebuilt by others in a different manner after due consultation; thus mistakes are corrected and improvements introduced. Interesting observations of this kind are recorded in the notebooks of Pierre Huber.*

Still more remarkable as architects are the termites or white ants of the tropics; they are also more advanced than the emmet in their social and political organization. In Africa their habitations begin with a series of pyramids about a foot high, which increase in size and number with the growth of the community, and are finally joined together and covered over with a cupola consisting of a firm coating of clay. The finished domelike structure often attains the height of ten and even twenty feet, and is made of clay, stones, pieces of wood, and similar materials cemented together with the mucilaginous spittle of the termites. The cone-shaped hillocks resemble haystacks, and at a distance are easily mistaken by travellers for the huts of the natives. Indeed, at first sight a termite village can hardly be distinguished from a negro village. They are so solidly built that they not only resist storms and the assaults of foes, but also sustain the weight of a man and do not yield to the pressure of a heavily laden wagon. It is said that gazelles, buffaloes, and even elephants are seen standing on them, and using them as points of observation.†

Some species of termites build in the form of truncated columns or gigantic fungi with a round roof projecting about two or three inches and resting on a

---

* Both François Huber and Darwin have noticed a like liability to error on the part of bees in building comb cells, thus proving that they are not always guided by unerring instinct. To err is apian as well as human, and the tendency is due in both cases to the same cause—namely, fallibility of the reasoning faculties.

† Büchner, pp. 226–228.

cylindrical base some four or five feet high. In regions subject to inundations their dwellings are constructed like barrels round the trunks of dead trees and communicate with the ground by means of corridors or passageways bored through the wood.

Blanchard, in his Rapport sur les Travaux Scientifiques des Départements en 1868, describes the elaborate interior arrangement of the termite mounds with their myriads of rooms, cells, nurseries, storehouses, sentry chambers, passages, halls, arcades, and other large or small spaces set apart for particular purposes and forming parts of a well-considered plan. In the centre is what he calls the "royal residence" with a high-arched ceiling resembling an old-fashioned oven. Here the royal pair dwell, or rather are kept captive, since the entrance is so narrow that it is impossible for them to go out and in. They are well fed and assiduously served by the workers, but never leave the apartment and are virtually prisoners of state, treated with respect, like the mysterious man in the iron mask, but nevertheless restrained of their liberty. The prolific queen assumes enormous dimensions, becoming two or three thousand times as large as an ordinary termite. Adjoining this lying-in room (for such is its essential character) are nurseries for rearing the young, chambers for servants or attendants of the queen, barracks for soldiers, closets or cupboards filled with gums, resin, dried juice of plants, seeds, fruits, and other edibles or condiments.

In the centre of the termite mound is a large space with passages leading to it from all sides, which, in the opinion of Bettzieh-Beta, serves the purpose of a forum or place of public meetings, held, as he assumes, for the discussion of questions of general interest. Others maintain that it is intended to promote ventilation.*

Dr. H. Hagen states that some termites, in order to get at a sack of meal standing on the floor and effectually protected against their direct encroachments, gnawed a hole through the ceiling just above the sack and built a tube downward through the air until they reached the object of their desire. But finding it impossible to carry the meal up through this straight and perpendicular passage, they constructed another by the side of it with a spiral ascent like that of the campanile of St. Mark's in Venice. By thus taking advantage of the principle of the inclined plane, which plays such an important part in modern engineering and roadmaking over mountains, they easily succeeded in securing the meal.

Blanchard also compares these insects to skilful engineers, and confirms the observations of other naturalists as regards the ability with which they design and construct tubular bridges from one point to another in the form of an arch or succession of arches. In the cellar of the prefecture of La Rochelle, in southern France, they made hollow columns as large as a thick straw from the ceiling to the floor, using them as lines of transit and transportation to the upper stories of the building. He adds that they always take the shortest cut to their destination even when working underground. They are, therefore, suspected of sending out explorers by night, who survey the ground and indicate by signs on the surface of the earth the direction of

---

* Büchner, p. 229.

the projected subterranean passage. This supposition is regarded by Büchner (p. 236) as highly probable.

Smeathman describes the effect of making a breach in the mound of the termites. Immediately the soldiers rush out in the greatest rage in obedience to a signal given by a single sentinel or officer, who first appears in order to ascertain the nature of the attack. If the disturbance is not renewed the soldiers retire and the workers reappear and begin to repair the damages, only a few of the former remaining stationed here and there as a guard. If the mound is again disturbed, the workers vanish and the soldiers come out in force. There seems to be an exact assignment of duties to each class, the workers never fighting and the soldiers never working.

The workers are undeveloped females and the soldiers may be undeveloped males, although this is by no means certain. At any rate they are both classified as sexless and are both blind, the lack of sight being supplemented by a delicate sense of touch, which, as they live in the dark, serves them better than vision. They are distinguished from each other chiefly by the form and armament of the head; that of the worker being round and smooth and provided with a mouth adapted to the elaboration of materials for building purposes, while that of the soldier is very large and armed with pincers, pikes, or tridents, and long jaws with saws or sabres serving as weapons for assault. The proportion of soldiers to workers in a mound is about one per cent, so that the standing armies are relatively much smaller than those which have so often been a burden to European powers. In other words, ninety-nine hundredths of the population devote themselves to industry and only one hundredth to arms, thus indicating an advanced state of termitic civilization.

The termites are as energetic and ingenious in their destructive as in their constructive labours. Perhaps the most dangerous of these creatures (at least so far as we know) is the *Termes lucifugus* or light-shunning termite, which was introduced into Europe on exotic plants imported from Brazil. In southern France, especially in La Rochelle, Rochefort, and Bordeaux, they have eaten up furniture, caused wooden buildings to collapse, and ships of war to fall to pieces. They enter the foot of a table through the floor and gradually eat out the whole inside of it, so that it has the appearance of being perfectly solid when it is only a mere shell, which the slightest shock or pressure causes to crumble into a heap of dust and splinters. In consuming the corks of bottles they always leave a thin layer at the lower end sufficient to prevent the wine from flowing out and submerging them. In South America, India, and Egypt they have been known to destroy whole villages and compel the inhabitants to migrate. Curiously enough in consuming a house they spare the principal pillars, whose destruction would cause the whole building to fall; or when they devour the inside of these pillars, they fill the hollow with clay, which hardens and renders them stronger than ever. This remarkable foresight contributes to their self-preservation. But how should the insects know on which pillars or columns the edifice mainly rests? *

Bastian † states that the cities of the termites in East

---

* Cf. Dr. Hagen, cited by Büchner, p. 241.
† Die Völker des Oestlichen Indiens.

India are as tall as a man, some of them being simple massive mounds, and others resembling a regular castle with battlements, pinnacles, and turrets. In the savannas of Senegambia in western Africa the countless hills of the termites furnish an admirable material, out of which the aborigines as well as the canny Scotch missionaries construct their houses. They are also made into ovens by being hollowed out and the interior plastered with loam. These structures are relatively much larger than any reared by man. A pyramid bearing the same proportions to the size of its builders would be at least three thousand feet high, and a subterranean canal would be three hundred feet in the clear. In comparison with them old Roman and modern American edifices and aqueducts are insignificant affairs.*

Even instincts, which seem firmly rooted and are regarded as characteristic of the class, are by no means so persistent as is commonly supposed. The individual inherits, but soon loses them if they are not brought into early exercise. A duck or gosling, if reared in the house until it is two or three months old, has no greater liking for the water than a chicken, and if thrown into a pond will scramble out, showing signs of great fear of the element to which its web-feet are particularly adapted. An artificially hatched chicken does not attach itself to a hen more than to any other animal, but follows its first associate, a child, a cat, or a dog.

Buffon denies that animals are susceptible of what he calls "the perfectibility of the species." "They are to-day," he says, "what they always have been, and

---

* For facts and authorities see Büchner's Aus dem Geistesleben der Thiere, Leipzig, 1896.

always will be, and nothing more; because, as their education is purely individual, they can only transmit to their young what they themselves have received from their parents. Man, on the other hand, inherits the culture of ages and gathers and conserves the wisdom of successive generations, and may thus profit by every advance of the race, and, in turn, aid in perfecting it more and more."

This assertion has been repeated by scientists of the old school as though it were an axiom of natural history, instead of an arrogant anthropocentric assumption refuted by scores of well-authenticated facts. The whole process of domestication, which is to the lower animals what civilization is to man, and the possibility of producing and propagating desirable qualities in the race, run counter to Buffon's theory. The value of a horse's pedigree depends upon the transmissibility of distinctive characteristics which were originally peculiar to some individual horse, idiosyncrasies which commended themselves to man as worthy of preservation, or such as in the natural struggle for existence would assert and propagate themselves.

If the descendants of blood-horses do not inherit the individual training of their sires, neither are the children of scholars or musicians born with a knowledge of books or the ability to play on musical instruments. What is inherited in both cases is some particular disposition or endowment, a superior aptitude for the things in which their progenitors excelled. Indeed, this heritage is handed down in horses with surer and steadier increase, or, at least, with smaller loss and depreciation than in human beings, since they are mated with sole reference to this result; and there is no room left for the play of personal fancy and caprice, or for social, sentimental, or pecuniary considerations, which exert a baneful influence upon marriage from a physiological point of view, and contribute to the deterioration of the race. This is strikingly perceptible in some portions of Europe, where the struggle for existence, and especially for high social position, is exceedingly intense, and a large dower suffices to cover up all mental and physical deficiencies in the bride.

The scientific swine-breeder keeps genealogical tables of his pigs, and is as jealous of any taint in a pure porcine strain as any prince of the blood is of plebeian contamination. In both cases the vitiation bars succession, the one condition of which is purity of lineage. It is by the selection not only of the finest stock, but also of the choicest individuals for breeding, that animals are "progressively improved" both bodily and intellectually. This is, perhaps, most clearly observable in hunting dogs and race horses, which have undergone quite remarkable modifications within the present century owing to the extraordinary pains taken to develop and perfect their peculiar characteristics. In some instances unusual births or freaks of nature are preserved, and by persistently propagating themselves form the starting point of new species. A striking example of this perpetuation of individual peculiarities is the short-legged and long-backed Ancon sheep, a comparatively recent product of Nature rendered permanent by the care of man. A pointer, greyhound, or collie inherits and transmits to its offspring not only race attributes, but also acquired aptitudes in the same manner and to the same degree as a human being does who is distinguished for some special faculty.

There are prodigies of dogs which do not beget prodigies of puppies, just as there are men of genius whose children are by no means eminent for their intellectual endowments.

If the conceptual world of the lower animals is limited and fragmentary, so is that of savages and of ignorant and uncultivated men, who live for the most part in the present and the immediate past, and have a relatively narrow range of thoughts and experiences. Long-lived animals, such as parrots, ravens, and elephants, have an advantage over short-lived animals in the development of intelligence. Civilized man, however, not only lives his own individual life, and profits, like other animals, from the wisdom of his parents and the influences of his environment, but also, by means of written records, lives the life of the race, of which he enjoys the selectest fruits garnered in history.

It must also be borne in mind that dogs are and always have been bred for special purposes, such as pointing, retrieving, running, watching, and biting, but not for general intelligence. Mr. Galton, who calls attention to this fact, suggests that it would be interesting as a psychological experiment to mate the cleverest dogs generation after generation, breeding and educating them solely for intellectual power and disregarding every other consideration.

In order to carry out this plan to perfection and to realize all the possibilities involved in such a comprehensive scheme, it would be necessary to devise some system of signs by which dogs would be able to communicate their ideas more fully and more clearly than they can do at present, both to each other and to man. That the invention of such a language is not impossible is evident from what has been already achieved in the training of dogs for exhibition, as well as from the extent to which they have learned to understand human speech by mere association with man. Prof. A. Graham Bell believes that they may be taught to pronounce words, and is now making scientific experiments in this direction. The same opinion was expressed two centuries ago by no less an authority than Leibnitz, who adduced some startling facts in support of it. The value of such a language as a means of enlarging the animal's sphere of thought and power of conception, and of giving a higher development to its intellectual faculties, is incalculable.

Every dog trained as a hunter or herder is a specialist, and is prized for one fine capacity attained in some degree at the expense of mental proportion and symmetry; in miscellaneous matters outside of his province he may be easily surpassed by any underbred and mongrel but many-sided village cur. Modern scholarship shows a like tendency to psychical alogotrophy or one-sided intellectual growth. As science deepens its researches, each department of investigation becomes more distinct, and the toiler in the mines of knowledge is forced to confine his labours to a single lode if he would exhaust the treasures it contains. He sees clearly so far as his lantern casts its rays; but all outside of this small luminous circle is dense darkness.

If a race of superior beings had taken charge of man's education for thousands of years and conducted it on the same principle as that which has guided us in domesticating and utilizing the lower animals, what maimed specimens of humanity would have been the result! Slavery has always tended to produce this effect; but the slave, however degraded his condition, speaks the same language as his master, thereby profiting from his intercourse with those who are placed over him, and sharing in the general progress of society more fully than any dumb animal could do.

The influence of domestication on the mental development of animals depends upon the purposes which the domesticator has in view. If he regards them merely as forms of food, and his sole aim is to increase the amount of their adipose tissue and edible substance and thus get the maximum of meat out of them, then domestication tends to stupefy them. The intellectual training of the pig would naturally diminish the quantity of lard it would produce. So far as man is concerned, this latter function is the chief end of the porker's existence, and it must not be *tried* and found wanting in this respect, whatever may be its mental deficiencies. It must be fat-bodied whether it be fat-witted or not, and the natural qualities which do not contribute to its gross weight and enhance its ultimate value as victuals are systematically discouraged and depressed.

In view of the treatment that the pig has received for centuries at the hands of man, it is remarkable that the animal has retained so much of its original cunning and love of cleanliness as it now possesses. That a creature so fond of bathing in pure running water should be condemned to a filthy sty is an act of unconscious cruelty discreditable to human discernment. If the sow that has been washed returns to her wallowing in the mire, it is as a last resort in hot weather; she would much prefer a clear pond or limpid stream if she could get access to it.

Being fed and protected by its owner in its domestic state, the hog no longer needs to exercise the faculties which were essential to the self-preservation of its wild progenitors. The stimulus arising from the struggle for existence ceases, and, as it is reared solely to be eaten, its association with man does not call forth any new powers. In China and Polynesia, where the dog is esteemed chiefly as food, it is a sluggish and stupid beast. On the other hand, the pig can be trained to hunt, and not only acquires great fondness for the sport, but also shows extraordinary sagacity in the pursuit of game. It has an uncommonly keen scent, and can be taught to point better than the pointer. Curiously enough, when the pig is used for hunting purposes, the dogs, usually so eager for the chase, sullenly retire from the field and refuse to associate with their bristly competitor in venery. Possibly the hereditary and ineradicable enmity between the dog and hog as domestic animals may be a survival of the fierce antipathy which is known to exist between the wolf and the wild boar. In Burmah the ringed snake is trained for the chase, and is especially serviceable in flushing jungle-cock, since the reptile can penetrate the thickest underbrush, where it would be impossible for a dog or a falcon to go.

The tamability of an animal is simply its capability of adapting itself to new relations in life, and depends partly on its mental endowments, but still more upon its moral character. It is quite as much a matter of temperament and social disposition as of quickness of understanding. The elephant, dog, and horse among quadrupeds, the beaver among rodents, and the daw and raven among birds, are, for this reason, most easily tamed, and show the most marked and rapid improvement in consequence of their daily intercourse with man. Intellectual acuteness without the social affections and kindred moral qualities rather resists than facilitates domestication. Of all domestic animals the cat was the most difficult to tame, and it needed the patience and persistence so strongly characteristic of the ancient Egyptians, sustained by religious superstition, in order to accomplish this result. Even now the cat, although extremely fond of its home and capable of considerable attachment to persons, has never been reduced to strict servitude and become the valet of man like the dog, but has always remained to a certain degree what it originally was, a prowling beast of prey.

Barking in dogs is a habit due to domestication. The wild dog never barks, but only howls, like the Himalayan buansu, or merely whines, like the East Indian colsum; and the domestic dog reverts from barking to howling when it relapses into its primitive state. Wagging the tail is another mode of expression which the dog has acquired through association with man. It is well known, too, that a dog which has been reared by a cat adopts many of the habits of its foster-mother, such as cleaning itself with its paw; by continuously pairing such dogs and rearing them under like influences it would be possible to produce a canine species with feline traits, which should become permanent and transmissible.

A recent writer, Dr. Leopold Schutz, professor in the theological seminary at Trier, who may be taken as an extreme representative of the old orthodox school of zoöpsychologists, maintains that animals do not think, reflect, form purposes, or act with premeditation of any kind, have no freedom, no choice, no emotional or intellectual life of their own, but that a higher power performs all these operations through them as cunning pieces of mechanism. The bird sings, according to this theory, without any personal pleasure or participation in its song; it sings at a certain time and can not help it, nor is it able to sing at any other time. The living cuckoo is as automatic as the wooden cuckoo of a Black Forest clock, and under the same mechanical compulsion to sing its song when the appointed hour arrives. Altum, in his book on bird life (Der Vogel und sein Leben, Münster, 1868), infers from the fact that a bird sings more in the pairing season than at other seasons of the year, that its song is a "natural necessity," in which it takes no individual pleasure. But this conclusion by no means follows from the premises. The song is a means to an end, and has for its final object sexual attraction and selection. One would surely not be justified in inferring that a woman who dresses well, chiefly in order to gratify her husband or her lover, finds no individual æsthetic satisfaction in a fine gown; or that a man goes a-wooing from "natural necessity," and gets no entertainment out of courtship.

Prof. Schutz's doctrine that animals are mere puppets, whose movements are determined by the direct intervention of higher powers, seems to have been derived from what is recorded of the relations of these creatures to holy men in the legends of the saints, rather than from a scientific study of the book of Nature; his point of view is not that of the zoöpsychologist, but that of the hagiologist.

The chief difficulty attending the investigation of

mental processes in animals is that they can not express themselves in human language and explain to us their thoughts and feelings and the motives underlying their conduct. We are thus liable to misinterpret their actions and deny them many endowments which they really possess, just as the first explorers of new countries fail to discover in savages ideas and conceptions which are afterward found to characterize them in a remarkable degree.

We have happily rid ourselves somewhat of the ethnocentric prepossessions which led the Greeks, and still lead the Chinese, to regard all other peoples as outside barbarians; but our perceptions are still obscured by anthropocentric prejudice which prevents us from fully appreciating the intelligence of the lower animals and recognising any psychical analogy between these humble kinsmen and our exalted selves.

CHAPTER VII.

IDEATION IN ANIMALS AND MEN.

Prantl's doctrine of "time sense" as a specifically human endowment. Weakness of this theory. Examples of time sense in animals. Their social instincts and moral sentiments. Standard of animal virtue according to Spinoza. Animal herd and savage tribe. Criminal procedure of animals against delinquent members of the herd or flock. Civic life of ants and bees. Anarchism and barbarism in the hives. Depraving influence of alcohol on bees. Agricultural ants. African driver ants. "Cattle-lifting" ants. Formican slaveholders. Volume of brain in hymenoptera. Long and helpless infancy of ants. Altruism in animals. Dr. McCook on honey ants. Darwin's experiment. Use of tools by animals. Mechanical skill of trapdoor spiders. Elephants as dam builders. Use of implements by crows and cormorants. Wine-making apes in China. Monkeys as miners. Use of fire as an index of civilization. The logical faculty in monkeys. The soko as a humorist. Idea of personal property in animals.

THE late Prof. von Prantl * takes the ground that the lower animals are endowed with moral and intellectual faculties, but adds: "They are destitute of any logical apprehension and power of abstraction; for while they comprehend objects and their optical, acous-

---

* In a paper on Reformgedanken zur Logik, read before the Royal Bavarian Academy of Sciences, and printed in its Proceedings for March 6, 1875.

tic, and other efficient qualities in a certain abiding manner, they have no conception of substance or attribute, of coexistence or succession. Animals perceive also an actual causal connection, and are therefore capable of drawing causative conclusions, reasoning forward and backward, from cause to effect and from effect to cause, but not capable of a logical deduction; they seek a cause, but not a logical ground or reason, and are, by virtue of such endowment, wary and cautious, but without foresight" (*behutsam und vorsichtig, aber ohne Voraussicht*). In other words, "animals think without logic, but not therefore illogically."

Again, "in order to formulate precisely the distinction between man and beast," he sums up this difference in the succinct statement, " man has time-sense." Beasts have "space-sense," or the "sensual perception of expansive being," but not "time-sense; that is to say, the brain activity of man is competent to comprehend also pure succession as such, and the pure intensity of occurrence in general."

In proof of this proposition Prantl states that "man can count." Even without the use of names or numerals "he can fix the succession of days by marks, or express the number of objects lying before him gesticulatively with his fingers." This "sense of continuity, denied to the whole world of lower animals," renders man "conscious of being the same in a later as in a former time," and thus endows him with "immutable ego-consciousness, or Kant's transcendental apperception." It enables him to look before and after, to bind together the past and the future, and thus to create law and order, domestic, social, and political institutions, ethics, art, religion, science, and history, and to make external things serve his purposes and supply his wants. "Man, and man only, fabricates weapons and tools, kindles fire, plants seeds in the earth, and is alone capable of self-renunciation and suicide." "By virtue of this continuity of his self-consciousness and his look into the future, he transforms the realities around him and makes them minister to his ideals." The sole and ultimate source of all these higher developments and ideal acquisitions of humanity, individual, social, political, industrial, and artistic, is to be sought in "the far-reaching and fundamental postulate that man is endowed with time-sense."

For this reason man alone is able to distinguish between the subjective and the objective, to conceive of the subject as an object, and to apprehend mathematical truths and relations, which are purely ideal, as real. "It would be ridiculous to ascribe mathematics to animals; nevertheless the labours of the bee and of the spider excite astonishment; but inasmuch as, with genuinely animal limitations, they always appear in a definite geometrical form, they show that they are not products of spontaneous mathematical thinking."

Prantl also denies that expressions of sorrow, remorse, or gratitude on the part of animals furnish any evidence that they act under the impulse of moral ideas, but interprets them as having reference to their own well-being or comfort. To talk of the "art-instinct of animals" is, he thinks, a mere confusion of terms, "since we demand of art that it shall realize an idea." Still, after all his metaphysical distinctions, he admits that the essential nature of man as distinguished from that of the beast is "only the result of a progressive upward evolution." If this conclusion be correct, and it is all that the most advanced zoöpsychologist has ever claimed, then the distances (*Abstände*) between man and beast are not impassable, and even "human speech" (*die menschliche Sprache*) is but a higher development of "animal utterance" (*die thierische Kundgebung*).

The weak point of these speculations concerning the mental powers of animals is that they are too exclusively metaphysical, constituting a logical and systematic exposition of conceptions or notions without that accurate and exhaustive observation of facts which no acuteness of analysis and no vigorous process of pure thinking can supply. Not only is Prantl ignorant of the habits and aptitudes of animals, denying them capacities which they are known to possess, but he is liable to an opposite error, equally fatal to his theories, in his tendency to ascribe to the human race as a whole faculties which are characteristic of man only in a high state of civilization. He ignores the savage and the boor, and compares beasts with the most cultivated and most highly developed human beings, overlooking the long period which man existed on the earth before he even learned how to chip flints.

As to the "ideal-sense," upon which Prantl lays peculiar stress, there are low tribes in which it is wholly wanting, and which are as destitute of historical annals as any herd of apes. How much knowledge of the past may be transmitted from generation to generation by tradition in a community of monkeys it is impossible to determine. The amount of information thus preserved and accumulated in simian hordes is probably very small and exceedingly vague, since even human hordes, not native to the countries they inhabit, soon lose all recollection of the early migrations of their ancestors, and all traditions concerning the cradle of their race. This is why savages always regard themselves as autochthones, even in cases in which it can be clearly proved that they are not aboriginal to the soil, and that their immigration is of comparatively recent date.

There is no reason to believe that "time-sense," which Prantl claims to be the exclusive attribute of man, and from which he derives the superior mental evolution and equipment of the human race, is wholly lacking in the lower animals. Every creature endowed with personal consciousness and memory must know that it is the same being to-day that it was yesterday, or, in other words, that it exists in time. The possession of this knowledge does not imply the possibility of indulging in philosophical reflections about it any more than the possession of thoughts necessarily involves the power of thinking about thoughts, although it would be rash to affirm that animals may not be capable of giving themselves up to meditation by recalling mental impressions and making them objects of thought.

Time-sense is very highly developed in domestic fowls and many wild birds, as well as in dogs, horses, and other mammals, which keep an accurate account of days of the week and hours of the day, and have, at least, a limited idea of numerical succession and logical sequence. A Polish artist, residing in Rome, had an exceedingly intelligent and faithful terrier, which, as he was obliged to go on a journey, he left with a friend, to whom the dog was strongly attached. Day and night the terrier went to the station to meet every train, carefully observing and remembering the time of their arrival, and never missing one. Meanwhile he became so depressed that he refused to eat, and would have died of starvation, if the friend had not telegraphed to his master to return at once if he wished to find the animal alive. Here we have a striking exhibition of time-sense as well as an example of all-absorbing affection and self-renunciation likely to result in suicide.

Love, gratitude, devotion, the sense of duty, and the spirit of self-sacrifice are proverbially strong in dogs, and only a "hard-shell" metaphysician, who neither knows nor cares anything about them, would venture to deny them all moral qualities, and to assert that they are governed solely by a regard for their own individual well-being. There are also many apparently well-authenticated instances of animals deliberately taking their own lives; and without too credulously accepting anecdotes of this sort, in which it is difficult to determine whether the creature was a *felo-de-se* or the victim of an accident, there is no pyschological reason for rejecting them as old wives' fables.

Scorpions and serpents are especially prone to sting themselves to death when kept in close confinement. Some naturalists maintain that these creatures grow crazy before committing the fatal act; but it is difficult to determine whether the wounds are self-inflicted for the purpose of putting an end to their existence or are the result of attempts to defend themselves against an imaginary enemy. Mr. Holden, of the Lick Observatory, reports the case of a rattlesnake, which, after several unsuccessful efforts to escape from captivity, bit itself to death "in a most deliberate manner." He is convinced that the suicide was intentional.

According to Spinoza, benevolence in animals consists in the exercise of friendly feelings toward their kind, and this is all that we have a right to demand of them. A good cat, for example, is a cat that is good to her kittens, however cruel she may be to birds and mice. Indeed, her goodness, from a feline as well as from a human point of view, is in direct proportion to her destructiveness of the smaller rodents. A like standard of virtue prevails among low races of men, and constitutes the highest ideal of tribal ethics. The best man among barbarians is the one who is most terrible to their foes, and can put the greatest number of them to death in the shortest time. Such manifestations of love of kin and love of country are only enlargements of self-love; and it is a long way from this primitive form of egotism to universal philanthropy, and to the still broader benevolence which Buddhism inculcates toward all sentient creatures. One is inclined to pardon the gruff cynicism of Dr. Johnson in denouncing patriotism as "the last refuge of scoundrels," when one sees how much individual selfishness finds a covert under this fine-sounding word, and what fierceness of interdynastic and international strife it is made to provoke and to palliate.

Not only the social instincts, but also the moral sentiments growing out of social relations, are common to man and to beast. It is evident that germs of moral ideas and perceptions of moral obligations enter into the conjugal unions of beasts, and impart a certain stability and sacredness to these ties. Many animals are strict monogamists, and have thus attained what Aryan civilization now generally accepts as the highest and purest form of sexual affection and association. With beasts, too, as with men, it is the male which scruples least at trangressing the monogamous principle, and makes light of this breach of fidelity, treating it as a pardonable peccadillo.

The mandarin duck is proverbial for conjugal faithfulness, and the Chinese are accustomed to carry a pair of these fowls in bridal processions, as an emblem of connubial love and an example of constancy for the newly wedded couple. Canaries are also characterized by the same virtue, and the attempt to force them into bigamy by keeping one male and two females in the same cage is uniformly destructive of domestic bliss, and frequently fatal to the young. Jealousies are quite sure to arise in consequence of a preference of the male for one of his mates; and the consort that feels aggrieved by marital neglect will take every opportunity to avenge herself by pecking and pestering her favoured rival, and destroying her nest with its contents of eggs or callow brood. Even the young which are reared under such circumstances are far inferior in beauty and vigour, as well as in numbers, to the offspring of a peaceful monogamous canary household.

Whether the family may be the originary nucleus of the tribe, or, as is more probable, may have been developed through a process of differentiation out of a primitive community, whose members lived in sexual promiscuity,* the impulse to herd, as well as the purposes it subserves, are the same in savages and in beasts. Wolves hunt in packs; cattle, horses, and sheep

---

* We may state that Westermarck, in The History of Human Marriage, takes a different view, but does not settle the question definitely.

unite for mutual protection; and this tendency remains even after their domestication, when it is no longer essential to their safety, and becomes, as in man, a purely social feeling. Birds of passage assemble for their annual or semiannual migrations, and separate into families as soon as they have reached their destination; still preserving, however, their larger and laxer social organization as "birds of a feather," which enables them to "flock together" again with facility, whenever the general interest requires united action of any kind. This sense of community is especially strong in rooks and storks, which seem to have a regular system of government, by means of which they enforce discipline, reproving and correcting deviations from their common standard of rectitude, and even inflicting capital punishment for certain transgressions. In such cases the family ceases to exercise jurisdiction over its own members, and recognises the superior penal authority of the commonwealth.

The instances recorded of animals holding courts of justice and laying penalties upon offenders are too numerous and well authenticated to admit of any doubt. This kind of criminal procedure has been observed particularly among rooks, ravens, storks, flamingoes, martins, sparrows, and occasionally among some gregarious quadrupeds. It is as clearly established as human testimony can establish anything that these creatures have a lively sense of what is lawful or allowable in the conduct of the individual, so far as it may affect the character of the flock or herd, and are quick to resent and punish any act of a single member that may disgrace or injure the community to which he belongs.

Sometimes an irascible husband may take the law into his own hands, and summarily avenge himself on a faithless wife and her guilty paramour without bringing the case before a general assembly of his kind. Usually, however, it is the whole body which, after due deliberation, pronounces and executes judgment and maintains the majesty of the law. The penalty does not always involve the forfeiture of life, but varies in rigour according to the turpitude of the offence; the culprit being often condemned to a severe castigation, after which he resumes his position in society a sadder and wiser member of it.

A general assembly of storks was held on June 25, 1896, on the meadows of Enzheim in Lower Alsatia, and remained in session more than forty-eight hours. On the first day there were one hundred and ninety-two storks present, and on the second day one hundred and eighty-nine. It was evidently a very lively conference and their chatter could be heard at a great distance, but there was nothing to indicate whether the meeting was simply a social reunion or whether the deliberations were of a judicial or political character. At any rate, it was not accidental, and clearly had some definite purpose in view.

Dr. Edmonson states that the hooded crows in the Shetland Islands hold regular assizes at stated periods, and usually in the same place. When there is a full docket, a week or more is spent in trying the cases; at other times, a single day suffices for the judicial proceedings. The capitally condemned are killed on the spot.

The owner of a house near Berlin found a single egg in the nest of a pair of storks, built on the chimney, and substituted for it a goose's egg, which in due time was hatched, and produced a gosling instead of the expected storkling. The male bird was thrown into the greatest excitement by this event, and finally flew away. The female, however, remained on the nest, and continued to care for the changeling as though it were her own offspring. On the morning of the fourth day the male reappeared accompanied by nearly five hundred storks, which held a mass meeting in an adjacent field. The assembly, we are informed, was addressed by several speakers, each orator posting himself on the same spot before beginning his harangue. These deliberations and discussions occupied nearly the entire forenoon, when suddenly the meeting broke up, and all the storks pounced upon the unfortunate female and her supposititious young one, killed them both, and, after destroying the polluted nest, took wing and departed, and were never seen there again.

It happens occasionally that the confidence of the male stork in the virtue of his spouse is too strong to be shaken even by the presence of such questionable progeny; or, if he suspects her of frailty, he deems it best to condone the fault. They then unite in exterminating the bastard brood, and prudently keep the mysterious episode of ciconian domestic life to themselves.

Prof. Carl Vogt tells the story of a pair of storks which had lived together for many years in a village near Soletta. One day, while the male was absent providing for his family, a younger suitor appeared, and began to pay court to the wife. She received his addresses at first with indifference; but as the woman who hesitates is lost, so she finally fell into the snares of her passionate and persistent adorer. His visits became more frequent, and at last he succeeded in so completely fascinating the matron that she was persuaded to accompany him to a marshy meadow, where her unsuspecting husband was engaged in catching frogs, and to join her gay paramour in putting the old stork to death.

A similar case occurred recently in north Germany. A pair of storks had had their nest on the roof of a barn for several seasons, without any apparent discord in their domestic relations. Suddenly, early in the spring, a powerful male stork made his appearance, and violently attacked the husband, who bravely defended himself; his spouse, strangely enough, taking no part in the fray. The assailant withdrew toward evening, his feathers dappled with blood, but renewed the attack on the following morning. The proprietor of the estate on which the scene took place resolved to interfere and shoot the intruder, but unfortunately aimed at the wrong bird and killed the husband. After this mishap, the female remained quietly perched on the roof by the side of the stranger, with whom she soon began to chatter in a very lively manner. The talk continued for about an hour, when both storks, as with one accord, fell upon the nest, threw out the eggs, tore it in pieces, and, after gazing for a moment on the ruins, rose together into the air, and, mounting in ever higher circles, vanished from view. Here the wife was at least accessory to the crime after its commission, and her conduct during the combat would seem to indicate that the strange stork was her accepted lover, and his coming preconcerted. Such occurrences, however, are exceptional. As a rule, storks are distinguished for conjugal fidelity no less than for their superior intelligence and the strong ties of affection which they form for human beings.

Ravens also have been known to destroy a nest in which a young owl had been discovered, and to kill both the birds whose home had thus suffered contamination, being evidently determined that the ancient and honourable race of *Corvus corax* should not be corrupted; and cocks, in several cases, are said to have killed hens which had hatched the eggs of ducks or partridges. One would hardly suspect such susceptibilities in a polygamous fowl, and least of all in our sultan of the barnyard, who guards his harem with the fierce jealousy of a Turk, but bears his paternal responsibilities very lightly, leaving the brooding mothers and their young for the most part to shift for themselves.

An unusually large number of ravens was recently observed on the trees in the Treptow Park of Berlin. They began to assemble about noon and continued to arrive from all points of the compass until three o'clock. After croaking together in loud tones for some time, they all pounced upon one bird sitting apart on a lower limb and belaboured it with their strong beaks until it was covered with blood and fell dead to the ground. Thereupon they all flew away in different directions. It is evident that this corvine convention was preconcerted, and that the purpose of it was to punish a guilty member of the community; but it was only after a thorough discussion of the matter that the sentence of death was passed upon the culprit and immediately executed.

O. Flügel, in his volume Das Seelenleben der Thiere (page 52), admits the truth of the stories about storks, but doubts whether their conduct under the circumstances has been correctly interpreted. "How do we know," he asks, "that the female stork is put to death as a punishment for adultery? . . . Perhaps she was sick and therefore unable to resist the suit of the male stork, and was killed because she was sick, as often happens in such cases. Perhaps, too, she was suffering, not exactly from a disease, but from a bodily abnormity." It is quite as difficult to imagine storks congregating for diagnostic as for judicial purposes. A ciconian medical consultation is not less marvellous and incredible than a ciconian court of justice. The assumption that she was physically infirm, instead of being a frail creature from a moral point of view, does not simplify the matter or render the proceedings of the storks more intelligible. If we begin to indulge in hypotheses of this sort, we may as well suppose that she was acting under the influence of hypnotic suggestion or some other irresistible form of fascination.

Indeed, in another passage (page 59), Flügel introduces hypnotism as a means of explaining actions on the part of animals, which might be far more satisfactorily explained by simply assuming that they are capable of practising deceit. Thus the opossum seeks to escape danger by feigning death, and the partridge pretends to be lame and limps off in an opposite direction in order to attract the attention of the pursuer to herself, and thus divert it from her young. "Perhaps," says Flügel, "fear makes the fowl really lame and throws the quadruped into an actual spasm or kind of hypnotic state." It is very queer that no amount of peril and terror should make the partridge go lame except when her young are with her;

and any one who has hunted opossums knows that in feigning death they show no symptoms of paralysis, but remain perfectly warm and limp, breathing naturally, though scarcely perceptibly. Herr Flügel is determined not to concede to the lower animals the exercise of rational faculties, and in order to avoid this necessity does not hesitate to give rein to the wildest conjectures.

As we have already seen, the impulses and motives which lead to the commission of crime are essentially the same in beasts and in man, and students of penal jurisprudence are just beginning to learn that the psychology of criminality in civilized society can never be fully understood except by a careful scientific study of it not only in savages, but also in the lower animals. The incentives to deeds of violence are pretty much the same in both. Many actions, such as the killing of deformed or sickly infants and of old and infirm individuals, are common to barbarians and to beasts, and are regarded as right because they contribute to the collective strength and consequent safety of the tribe or herd; but with the civilization of man and the domestication of the brute this precaution is no longer needed, and the primitive practice is abandoned. Mice take excellent care of their aged, blind, or otherwise helpless kin, concealing them in safe places and providing them with food. It must be remembered, however, that the mouse has lived in a semidomestic state as the companion of man from time immemorial.

In the development and organization of social and civic life the bee and the ant hold the foremost place among articulates, corresponding to that of man among vertebrates. They stand respectively at the head of their class, and represent the highest point attained by insect and mammal in the process of evolution. As regards form of government, it is a mistake to speak of the bee state as a monarchy; it is, on the contrary, the most radical of republics, or rather a democracy of the most rigorous kind, with absolute power vested in the working class. The claims of "labour" to the exercise of supreme control in political affairs are here fully recognised and practically realized. The so-called queen is really the mother of the hive; her functions are maternal rather than regal. If she may be said to reign in a certain sense, the workers rule, deciding all questions and performing all acts affecting the common weal. The existence of but a single queen leaves no room for those dynastic enmities and rivalries which have so often disturbed the peace of human empires, and inflicted such untold misery upon mankind. If perchance two queens are produced at the same time, instead of forming factions in the state and exciting civil war, they contend personally for sovereignty, until one of them is killed. Sometimes the workers intervene, and put the less desirable of the claimants to death; or if the hive is populous and circumstances are favourable, a portion of the inmates swarm and carry off one of the contestants to found a new colony. In all these operations the queen initiates nothing; she is a passive instrument in the hands of the workers, whose decisions she accepts, but does not influence in the slightest degree. There is no "blue blood" in her veins except such as may be produced by a process of pampering; she is simply a worker, taken in a larval state and fattened into regal favour and function by what Huber calls "royal treatment;" that is, by relieving her from all toil and supplying her with richer nutriment. If, on account of bad weather or for any other reason, the bees do not wish to swarm, they do not hesitate to throw all superfluous members of the royal family out of the hive. The institution of appanage is unknown to apian communities. But, in order to provide for emergencies, several larvæ are reared each in a single cell, which the old queen is never permitted to approach, since she is as jealous of these royal scions as was ever Persian padishah of his next of kin. For this reason they are kept in close confinement until they are needed.

The conjugal relation of the queen to the drones is polyandrous. "Her male harem," says Büchner, "is larger than the female harem of any Oriental despot, and consists often of six to eight hundred drones, who are for the most part utterly useless members of the community, since a single drone suffices to impregnate the queen, and the drones neither work nor are they armed with stings with which to defend the hive. They constitute a sort of hereditary peerage, letting themselves be served by the industrious workers and contributing nothing directly to the promotion of the common weal, but leading a lazy and comfortable life of leisure and pleasure from May to August, free from care and from toil, and doubtless with no presentiment of the fearful fate awaiting them in the autumn of their existence." This superfluity of drones, so opposed to the wise economy of Nature, is regarded by Büchner as the survival of a period when the bees lived in small, independent colonies and the drones, as they flew abroad a-wooing, were exposed to greater dangers than at present. The gathering of bees into hives under the care of man has entirely changed this condition of things, and thus left an excess of drones far above the number necessary to the propagation of the race.

Doubtless the queen has certain constitutional rights, but they are very limited. She is in the condition of Queen Victoria with Mr. Gladstone as prime minister: she is not asked what ought to be done, but is simply told what the cabinet intends to do, and is expected to indorse it, whether agreeable to her feelings or not. But this relation does not prevent a strong sentiment of loyalty toward her on the part of the workers, who are ready to defend her at the risk of their own lives.

On the other hand, they do not show the slightest affection for the males, or drones, who are in the unenviable position of prince consorts, or mere propagators of the race. No provision is made for them when the winter supplies of food are laid in; they fulfil their mission in summer, flying abroad on wedding tours with the queens of various hives and enjoying their honeymoon; but with the early frosts they are thrust out of the hives, and perish of hunger and cold. Meanwhile the queens preserve the sperm in a sac, and use it at pleasure for fecundating the eggs; as the fecundated eggs produce females and the unfecundated males, the numerical relation of the sexes can be easily regulated. The workers, or neuters, are really females, whose sexual organs remain rudimentary because all their energies are absorbed in labour. The ovary is only partially formed, and they are incapable of laying eggs; but it needs only a course of "royal treatment," consisting of luxury and idleness, to develop any of the larvæ into queens. It is asserted that workers do sometimes, though rarely, lay eggs; but this capability implies some degree of development beyond the condition of a "neuter" in the strict sense of the term, a change which might easily escape the eye of the "practical apiarist," who is usually less interested in the habits of bees than in the market price of honey. The queen has no heirs, either apparent or presumptive, and no right of succession is recognised. Any larviform worker can be metamorphosed into a queen, as every American schoolboy is a possible President of the United States.

That this perfect social and industrial organization, in which the principle of the division of labour is so admirably applied and a career opened to every talent, is the result of gradual growth and evolution is evident from the more primitive habits of other hymenoptera, such as wasps, hornets, and humblebees. Tame honey-bees also differ greatly in this respect from wild ones, and are known to have changed their manner of life and to have improved their methods of work to a considerable extent within the memory of man. A German writer states that when the European bee was imported to Australia, after a few years' experience of perpetual summer, it ceased to lay up winter stores of honey, making only what it wanted to eat from day to day. This fact, less edifying to the practical apiarist than instructive to the zoöpsychologist, furnishes the basis of Rosegger's charming tale and socialistic satire of anarchism in the hives. Here we have an example of radical changes in the habits of bees within the memory of the present generation as the result of new climatic conditions and the modifying influence of environment. This phenomenon is wholly inconsistent with the assumption of an innate and irresistible labour instinct.

Populous and powerful bee communities sometimes relapse into barbarism, renounce the life of peaceful industry for which they have become proverbial, acquire predatory habits, and roam about the country as freebooters, plundering the smaller and weaker hives, and subsisting on the spoils. These brigand bees seldom reform: if they busily "improve each shining hour," it is not to " gather honey all the day from every opening flower," but to range the fields in looting parties, and ransack the homes of honest honey-makers. Weygandt (in the periodical Die Biene, 1877, No. 1, cited by Büchner: Aus dem Geistesleben der Thiere, p. 322) describes the evolution and exploits of these "milking bees," as he calls them, and shows how the successful raid of a few individual filibusters soon converts the whole hive into a lawless band of depredators, who live by plunder. Against these anarchists of apian society and other foes the honeybees often fortify their hives, barricading the entrance by a thick wall, with bastions, casemates, and deep, narrow gateways. When there seems to be no immediate danger of hostile attack, these defensive works, which seriously interfere with the ordinary industrial life of the hive, are removed, and not rebuilt until there is fresh occasion for alarm. Jesse (Gleanings in Natural History, i., 21) states that the bees of one of his hives built a regularly constructed fortress wall before the entrance, in order to defend themselves more effectually against the raids of wasps, and adds that a small number of bees were thereby enabled to ward off these foes. The Swiss naturalist Huber, who began to publish his observations on the habits of bees more than a century ago and is generally recognised as a high authority, noticed that his bees erected a wall against the inroads of the death's-head moth in the spring of 1804, but removed it in the spring of 1805, in which year no moths of this kind were seen. In 1807, when the moths reappeared in considerable numbers, the bees again fortified the entrance and kept it in a state of defence till 1808. These barricades were made of propolis. Büchner (page 308) records other cases of a similar character in Hungary, where the death's-head moth (*Sphinx atropos*) frequently molests the hives. The common bee (*Apis mellifica*) not only rifles the nest of the bumblebee (*Bombus*), but numbers of them often surround one of the latter and force him to give up the sacs of honey he has gathered. The clumsy and not very courageous bumblebee submits to the demands of these highwaymen, surrenders his treasure without much ado, and then flies afield in search of more.

Büchner states that honest and industrious bees degenerate into vagabonds and robbers through the use of alcohol. If they are fed with a mixture of honey and brandy, they become passionately fond of it, get habitually drunk and disorderly, and gradually cease to work. The pangs of hunger, the penalty of one vice, drive them into another, and they take to theft and freebootery, as men do under similar circumstances. Instinct is not strong enough to resist the depraving influence of intoxicating liquors and save them from a downward career of demoralization and criminality.

According to recent observations made by Mr. Lawson Tait, wasps get drunk on the juice of plums, grapes, and other fruits, which is converted into alcohol by the process of decay. While in a state of intoxication, their sting is uncommonly venomous and produces the symptoms of nerve poisoning. The wasps become so addicted to this fermented juice that they take to it again as soon as they have slept off their drunkenness, and thus pass in rapid succession from one paroxysm of inebriety to another. Hens, too, which have access to the refuse of distilleries, soon become habitual drunkards, stop laying eggs, and show no desire to rear broods of chickens. In December, 1896, an owner of poultry in England brought a suit for damages against the proprietor of a distillery, because owing to the corruption of a neighbouring stream his fowls had become hopeless inebriates and thereby wholly worthless.

It is undeniable that, in the life of the honeybee, a sort of historical connection exists between the mother-hive and her colonies. This sense of kinship extends to the colonies of colonies, and thus gives rise to something like international relations between a large number of apian communities, which share the friendships and the hatreds of the original stock and transmit them to their posterity. Lenz relates his own experience on this point. Six of his hives were blown down by the wind; he hastened to set them up again, but the bees, rushing out and seeing him thus engaged, regarded him as the cause of the disaster, and stung him. For years afterward they pursued him whenever he approached their hives, and this unjust antipathy was inherited by all the swarms which issued from these hives and founded colonies elsewhere.

Here we have a striking instance of hereditary enmity, such as often characterizes families, tribes, and clans, and takes the form of the vendetta. The bees that had suffered the supposed wrong never forget it, and communicated their feelings to their descendants by way of tradition.*

In order to test the intelligence and foresight of bees Huber put them into a hive with a glass floor and a glass ceiling. It is well known that bees have great difficulty in building their cells on glass on account of its smooth surface, and avoid doing so if possible. In the present case they began to build their comb, not from the top or the bottom, as usual, but on one of the perpendicular sides over toward the opposite wall, but before they had reached it Huber substituted for it a glass plate. The result was that they ceased building in that direction, and, turning at a right angle, extended their comb to the other wooden side and fastened it there. They did not wait until they had come in contact with the glass before changing their plan, but foresaw and avoided the difficulty in the manner described.

Prantl's assertion that animals do not plant seeds in the earth and raise crops is merely one of many *a priori* deductions from his assumption that they lack time-sense, and therefore can have no appreciation of the succession of seasons. All facts opposed to this inference he would treat with a sceptical shrug of the shoulders, or relegate with an incredulous smile to the realm of fable. Nevertheless it is only by the careful observation and critical sifting of facts that such questions can be decided.

It has now been ascertained beyond a doubt that in

---

* Cf. Wundt, Vorlesungen über die Menschen- und Thierseele, ii, 196-200. Also article Bees in Encyclopædia Britannica.

Texas and South America, as well as in southern Europe, India, and Africa, there are ants which not only have a military organization and wage systematic warfare, but also keep slaves and carry on agricultural pursuits. Nineteen species of ants with these habits have been already discovered, and their modes of life more or less fully described.

Nearly half a century ago Dr. Linsecom began his studies of the Texan agricultural ant (*Atta malefaciens*), and after devoting some fourteen years to this subject communicated the results of his researches to Mr. Darwin, who embodied them in a paper read before the Linnean Society of London, April 18, 1861. This ant, he informs us, "dwells in what may be termed paved cities, and, like a thrifty, diligent, provident farmer, makes suitable and timely arrangements for the changing seasons. . . . It bores a hole, around which it raises the surface three and sometimes six inches, forming a low circular mound having a very gentle inclination from the centre to the outer border, which, on an average, is three or four feet from the entrance. On low, flat, wet land, liable to inundation, though the ground may be perfectly dry at the time when the ant sets to work, it nevertheless elevates the mound in the form of a pretty sharp cone to the height of fifteen to twenty inches or more, and makes the entrance near the summit. Around this mound, in either case, the ant clears the ground of all obstructions, and levels and smooths the surface to the distance of three or four feet from the gate of the city, giving it the appearance of a handsome pavement, as it really is. Within this paved area not a blade of anything is allowed to grow, except a single species of grain-bearing grass. Having planted this crop in a circle around, and two or three feet from the centre of the mound, the insect tends and cultivates it with constant care; cutting away all other grasses and weeds that may spring up among it, and all around outside the farm circle to the extent of one or two feet or more. The cultivated grass grows luxuriantly, and produces a heavy crop of small, white, flinty seeds, which under the microscope very closely resemble ordinary rice. When ripe, it is carefully harvested, and carried by the workers, chaff and all, into the granary cells, where it is divested of the chaff and packed away. The chaff is taken out and thrown beyond the limits of the paved area. During protracted wet weather, it sometimes happens that the provision stores become damp, and are liable to sprout and spoil. In this case, on the first fine day, the ants bring out the damp and damaged grain, and expose it to the sun till it is dry, when they carry back and pack away all the sound seeds, leaving those that had sprouted to waste." They also check the tendency of the seeds to germinate by biting off the incipient sprouts, treating them as a farmer does his potatoes or onions under similar circumstances.

In pasture lands, the grass cultivated by the ants is liable to be cropped by cattle, and thus prevented from bearing seeds and producing a harvest. In order to avert such a disaster, the ants avoid the meadows, which are given up to grazing, and establish themselves in the fence corners of cultivated fields, along garden walks or near gateways, or in other protected places, where their crops run the least risk of being destroyed.

These observations, the truth of which is amply confirmed by other writers, as, for example, by Dr.

Henry C. McCook in The Agricultural Ants of Texas, are a complete refutation of Prantl's zöopsychology; for no husbandman ever showed greater skill in adapting himself to circumstances, or manifested a higher degree of intelligence and foresight in conducting his agricultural operations, and in consulting for this purpose the nature of the soil and the variety of the seasons, than are exhibited by these marvellous insects.

Indeed, nearly all the institutions and gradations of culture and civilization which the human race has passed through, and of which we find survivals among the different tribes of men, exist also among ants. Besides the tillers of the soil just mentioned, there are other species, like the Peruvian cazadores, which still lead a nomadic life, having no permanent homes, but wandering from place to place; entering the houses of the natives by millions; killing rats, mice, snakes, and all sorts of vermin; devouring offal; and performing in general the useful functions of itinerant scavengers. On the approach of these hordes the inhabitants quit their dwellings, and do not return until the invading host has passed on. Dr. Hans Meyer, in an account of his ascent of the Kilima-Njaro, in equatorial Africa, states that his camp was one night attacked by an army of driver ants, and had to be abandoned. He describes the army as divided into three distinct classes, or castes, superior officers, underofficers, and the rank and file, each of which is provided with mandibles of different size and efficiency as weapons, and corresponding with the duties they have to perform. Other ants have advanced beyond this nomadic life of pillage, and have acquired fixed habitations; they do not cultivate the soil, but keep herds of aphides, or plant lice, which yield them a milky substance, and are also slaughtered for food.

The desire to get possession of these aphides is often the occasion of fierce raids of one community of ants upon another, forcibly recalling the cattle-lifting forays of Scotch clans, once so common on the northern border of England. A recent observer, Mr. James Weir, Jr., gives a graphic description of a predatory incursion of this kind made by an army of black ants (*Lasius niger*) into the domain of some yellow ants (*Lasius flavus*), whose herd of aphides was feeding under guard. The invading myrmidons "were marching in full battle array, with a skirmish line in advance. They came on with a rush, as if they intended a surprise. Some outposts, or pickets, of *Lasius flavus* discovered them when ten or twelve feet away from the town of *Lasius flavus*. These pickets raced home and gave the alarm. Immediately the inhabitants poured out and arranged themselves in front of their beloved herd. Skirmishers were thrown out and soon met the advancing *Lasius niger*. In a few moments the battle was on, and it was a battle to the death. The *Lasius niger* outnumbered the *Lasius flavus* three to one. As near as I could reckon there were about fifteen hundred of the blacks and about five hundred of the yellow ants. The yellow ants were larger and stronger, but the blacks were more agile. The yellow *Lasius* rushed at her enemy with open mandibles, and seizing her by the middle, crushed her through and through. The black *Lasius* endeavoured to get behind her enemy and then seize her by one of her legs. If she succeeded in her attempt, no bulldog ever held on with greater tenacity. As soon as possible another black ant would come to her assistance, and mounting on the back of the yellow ant would begin at once to gnaw through the thoracic wall. In a few seconds the shell would be eaten through, the vitals would be reached, and the yellow ant would sink down in the struggle of death. Not until certain that she was dead would *Lasius niger*, who had her by the leg, loosen her hold. *Lasius niger*, in this foray, came in light marching order. They carried no commissariat department, no ambulance corps. *Lasius flavus*, on the contrary, had both. When wearied or wounded the yellow ants would drop to the rear and communicate their wants. The ambulance corps dressed their wounds with their tongues; the commissariat refreshed them by regurgitating food into their open jaws. All through the battle I noticed this wonderful power of intelligent communication. *Lasius flavus* sent repeatedly back to the town to bring out the stragglers. It was like a well-ordered battle between human beings. These ants acted as though governed by an intelligence analogous to that which directs the actions of men. In the end, *Lasius niger* won the victory, but not until they had killed every *Lasius flavus*, and lost two thirds of their own number. The survivors carried off the bone of contention, the herd of aphides, to their own nest, some fifty feet away."

The slaveholding ants are of several kinds, and differ greatly in the manner in which they treat their vassals. Some make them do all the work under the direction of overseers; others share their labours; while still others have fallen into such habits of luxury as to be unable or unwilling to wait upon or even to feed themselves, and are carried about and provided with food by their body servants. In many cases this sybaritism is the mere ostentatious love of being served. The incapacity is not physical, but moral, and arises from an aristocratic aversion to any kind of menial labour, from the pleasure of being served by a train of obsequious attendants, and the notion that it is more dignified and distinguished to be borne along and to have food put into their mouths than to walk on their own legs and to help themselves to victuals; since these apparently so helpless ants are agile and energetic enough as warriors, when it is a question of conquering and plundering their peaceful neighbours. It is the false sense of honour, fostered by the military spirit, which takes pride in brandishing a sword and, on the slightest provocation, plunging it into the vitals of a fellow-man, but would deem it a deep disgrace for an officer to brush his own clothes or black his own boots.

Sometimes, in consequence of severe exactions, the slaves rise in revolt, and are mercilessly put to death; and formican like old Roman law seems to recognise the right of the master to inflict summary capital punishment in such cases. This power is often exercised by the red-bearded ant (*Formica rubibarbis*), who is a fierce slaveholder, and as pitiless in suppressing mutiny as was Barbarossa after the siege of Milan.

Ants differ in quickness of apprehension and in ingenuity quite as much as men do. Some with which Sir John Lubbock experimented, when cut off from their supply of food by the removal of a little strip of paper which had served as a bridge over a chasm a third of an inch in breadth, did not know enough to replace it. In similar cases, ants have been observed bringing straws from a distance for the express purpose of bridging chasms that separated them from a desirable article of food. Bridges for this purpose are often an inch long, and made of mortar or cement consisting of a mixture of fine sand with a salivary secretion.

In a monastery near Botzen, in the Tyrol, one of the monks put some pounded sugar, together with a few ants taken from an ant hill in the garden, into an old inkstand, which he suspended by a string from the crosspiece of his window. Very soon the ants began to carry the sugar along the string to their home in the garden, and returned with many others that went to work in the same way. After two days, although the greater part of the sugar was still in the inkstand, no ants were seen on the string; and, on closer examination, it was found that about a dozen of them were in the inkstand, busily engaged in throwing the sugar down upon the window sill below, where others were carrying it off to the hill. They thus saved themselves the trouble of climbing the whole length of the window and down the string into the inkstand and back again with their burdens, and avoided by this means an immense expenditure of strength and loss of time. This change in the plan of operations shows remarkable powers of observation and reflection, and was doubtless suggested by some of the more thoughtful and practical members of the community, and, after being communicated to the others, was adopted by them.

The intelligence of hymenoptera (ants and bees), like that of human beings, depends upon the development of the nervous system, and especially upon the size and structure of the brain. According to the tables published by Dr. Vitus Graber, the cerebrum of the bee's brain is $\frac{1}{180}$ and the cerebellum $\frac{1}{1000}$ part of its body, while the corresponding portions of the ant's brain are $\frac{1}{180}$ and $\frac{1}{800}$ of the size of the whole body. On the other hand, the brain of the May bug forms $\frac{1}{3000}$ and that of the water beetle only $\frac{1}{4200}$ part of its body. These fractions express approximately the relative mental capacity of each of the aforesaid insects, and the proportion is nearly the same as that existing between man and the larger mammals, such as the horse and the ox. The brain of the ant is, on an average, about one quarter the size of an ordinary pin's head, although it differs with different species. It is, doubtless, as Darwin has observed, the most marvellous physical atom in any living organism, not even excepting the brain of man, and shows what an amount of mental activity and energy may emanate from an exceedingly minute particle of nerve substance or be concentrated in the smallest ganglion. But the superior intelligence of ants, bees, and other hymenoptera living in highly organized communities, is due not only to the greater relative size, but still more to the complicated formation and composition of the brain, which is divided into two hemispheres, and differs from that of all other insects in its pedunculate character. The effects of injuries to the brain of an ant are analogous to those caused by injuries to the human brain: spasms, stupefaction, nervous prostration, the substitution of indefinite reflex action for voluntary movements of the body, raving madness, etc., according to the parts affected. Such phenomena have been often observed as the results of wounds received in battle, or in defending the larvæ or nymphs against the attempts of predatory Amazon ants to capture them. It is a curious and significant fact, too, that the babyhood of the ant is relatively quite as long as that of man, and during this plastic period of infancy the young of the genus *Formica* are quite as helpless and dependent upon the fostering care of their elders as are the young of the genus *Homo*. "The larvæ," says Büchner, "are occasionally sorted and divided into different groups, according to their age and size, so that one is involuntarily reminded of a school with distribution of the pupils into classes." "Nothing is more attractive," observes Blanchard, "than the incessant care of the ants for their larvæ. They keep them perfectly clean by rubbing and brushing them with their labial feelers, carry them in the morning to the upper stories of the nest, where it is warmer, and take them below again later in the day in order to escape the scorching rays of the midday sun. This transportation occurs as often as atmospheric changes and variations of the temperature require it. The soft bodies of the larvæ are borne between the firm jaws of the ants, but no injury or accident has ever been noticed; they are never bruised or wounded or hit against the hard walls." After reaching a certain stage of growth, during the course of the summer or sometimes not until the following spring, the larvæ spin themselves into so-called pupæ or chrysalides, popularly but falsely supposed to be ant's eggs and much in quest as food for caged birds. These pupæ or nymphs do not require feeding, but are nevertheless solicitously looked after by the working ants, licked, cleaned, carried about, and on fine days exposed to the air and light in front of the nest. When the sun gets too hot the attendants summon the workers, who carry the large, white, unshapely things back into the nest just as a cat carries her kittens. As soon as the pupæ have developed into ants they try to free themselves from their fibrous incasement, but seldom succeed without the aid of the workers, who unloose the web with their jaws and draw the young out of their place of confinement. After their release the skin, which still covers them like a shirt, is removed and their process of education begins. They are conducted through the nest and shown how to work. At first light tasks are assigned to them, such as taking care of larvæ, just as in human families older children are made to look after the infants. It is easy to distinguish the young ants from the old ones by their lighter colour, and thus to observe their actions. In a short time, however, generally in three or four days, they are fitted to perform the duties of a full-grown ant.

What Solomon says of the ants, that they have "no guide, overseer, or ruler" (Prov. vi, 7), is confirmed by modern entomologists, at least in its application to the ordinary species. No one ant seems to command the others, and Huber affirms that even the slaves are not subject to the slightest compulsion. "It is the consciousness of duty alone that preserves order and secures diligence." Forel asserts that the allusion to chiefs made by some writers (e. g., Ebrard) is "a mere figment of the imagination." If the larger and stronger ants take the lead in marching against foes, this prominence is due to their greater energy and efficiency as fighters and does not imply any other superiority. Even the warriors, who in some European and nearly all tropical species of ants appear to form a distinct caste, "never play an imperious part, but only serve the commonwealth." Under such circumstances a *coup d'état* or other arrogation of sovereignty by a successful soldier would be impossible.*

As regards moral attributes, says Dr. McCook in his work on the honey ants: "I am much inclined to the view that anything like individual benevolence, as distinguished from tribal or communal benevolence, does not exist. The apparent special cases of beneficence, outside the instinctive actions which lie within the lines of formicary routine, are so rare and so doubtful as to their cause that, however loath, I must decide against anything like a personal benevolent character on the part of my honey ants." †

It is often quite impossible to determine whether human actions arise from public spirit or private feeling; and an attempt to fathom the motives of ants, and to decide whether they are animated by a love of their kind and a desire to promote the general weal, or by a special good will toward individuals and what we call personal kindness, is attended with equal difficulty. But what the author affirms of honey ants is also true of savages, whose benevolence is tribal rather than personal; even civilized man, with rare exceptions, moves in the same narrow traditional rut, and is swayed in all his sentiments by national prejudices and prepossessions. The feeling of kinship is nevertheless especially strong in ants, and is not weakened by long absence. Mr. Darwin shut several of them in a bottle with asafœtida, and then released them and brought them back to their colony. At first their

---

* Cf. Büchner, pp. 54–57, 75–81.

† The Honey Ants of the Garden of the Gods, and the Occident Ants of the American Plains, page 45.

fellow-ants threatened to attack them and thrust them out, but soon recognised them under their offensive disguise, and received them with evident marks of affection. Still, no one would be justified in asserting that the elements of individual love and personal preference do not also enter into these relations. There is no doubt that strong attachments are formed between animals, and that they are capable of emotions of pity and acts of generosity not only toward their own kind, but even toward creatures of another species. A gentleman who had a great number of doves used to feed them near the barn; at such times not only chickens and sparrows, but also rats, were accustomed to come and share the meal. One day he saw a large rat fill its cheeks with kernels of corn and run to the coach-house, repeating this performance several times. On going thither he found a lame dove eating the corn which the rat had brought. Such an action on the part of human beings would be looked upon as a charitable desire to relieve the necessities of a helpless cripple, and every one would be satisfied with this simple explanation; but as a rat is assumed to be incapable of similar feelings, its conduct is regarded as the resultant of a series of impulses of sensation, perception, and conception, under which the animal is led to do wonderful things in an automatic way, without any consciousness of the purpose for which it does them; and thus a moral virtue is obscured and wholly hidden from view by a mass of metaphysical jargon.

A writer in the Revue d'Anthropologie relates the following story: The owner of a vegetable garden was surprised at the mysterious disappearance of carrots from a basket and asked the gardener what had become of them. The latter replied that he did not know, but would try to discover the thief. He accordingly hid behind the hedge, and had not waited long before the house dog came and carried off a carrot toward the stable, giving it to one of the horses, and wagging his tail with delight as his equine friend consumed it. The gardener was angry and, seizing a stick, was about to punish the pilferer for his excessive and rather eccentric exhibition of generosity, but the owner prevented him and secretly watched the dog, who continued to run to and fro between the garden and the stall until the entire stock of carrots was exhausted. Meanwhile the dog never bestowed a look, much less a carrot, on the horse in the next stall, who would have gladly eaten a share of the stolen fodder. Here we have a marked instance of altruistic sentiment manifesting itself in faithful friendship and even gross favouritism. That the lower animals are capable of feeling compassion and exercising charity toward creatures of their own or of other species is proved by numerous and well-authenticated examples of cats and dogs carrying food to other cats and dogs that were utter strangers to them, but were evidently suffering from hunger. In such cases the act of kindness does not even have its source in personal attachment, but springs solely from the purer fountain of pity and disinterested benevolence, and contains hardly a trace of what Spencer calls "ego-altruistic sentiment," self-gratification being wholly merged in the gratification of others.

Again the ability to use tools and to wield weapons, which Prantl derives from the possession of time-sense, is not exclusively human. Ants build bridges with splinters of wood, small pebbles, grains of sand, and other available materials, and tunnel small streams, and their skill in performing such feats of engineering and in meeting any emergencies that may arise is almost incredible; but the testimony of Bates and Bar and other naturalists leaves no doubt as to the reality of these achievements. They also make a clever and effective use of implements in capturing and killing the ferocious sand hornet, which they seize by the legs and fasten to the ground by means of sticks and stones, and then devour at their leisure. Here we have an unmistakable instance of the use of instruments for the accomplishment of a particular purpose. The same is true of the ant-lion when it prepares a pitfall and lies in wait for its prey, just as any hunter would do.

According to Moggridge, the antennæ of the trap-door spider (*Mygale fodiens*) are provided with a kind of rake, and the feet have prongs resembling the teeth of a comb. With the help of these instruments it digs subterranean tunnels or galleries, which it tapestries with a very fine silken web. The door, which closes the entrance, is very ingeniously constructed out of earth, woven firmly together with cobwebs. This door is very thick and broader above than below, so as to fit into the hole as a cork does into the mouth of a bottle. Its upper surface has the same colour as the surrounding earth, and it is therefore not easily discoverable. The hinge is made out of quite thick and firm fibres of silk, and the lock consists of a series of small holes in which the spider can stick its claws and hold the door fast from the inside. In going out the spider lets the door fall to, and lifts it up again in order to re-enter. This subterranean habitation, says Moggridge, is as far superior to that of the ordinary earth spider as the Mont

Cenis tunnel is to a common ditch. Erber, who studied the habits of the trapdoor spiders on the island of Tinos in the Grecian Archipelago, saw them spread nets after dark before their doors in order to catch night moths. In the morning these nets were removed. He also speaks of the marvellous skill and adaptation to circumstances with which they repair any injuries done to the doors or snares.*

Mr. Romanes seems to think that the only tool-using vertebrates are apes and elephants, but such a restriction is hardly justified by facts. The following incident, which is vouched for by Mr. William B. Smith, on whose farm at Mount Lookout it occurred, proves that an ass may understand the worth of weapons, and be able to avail himself of them. A donkey, which was in the same pasture with an Alderney bull, was frequently attacked by the latter, and worsted in the combat. Convinced that his heels were no match for his adversary's horns, the ass took a pole between his teeth, and, whirling it about, whacked his assailant so vigorously over the head that the latter was finally glad to give up the contest, and lived thenceforth on a peaceful footing with his long-eared and long-headed companion.

Cats and dogs open doors by pressing the latchkey, or cause them to be opened by pulling the bell cord or lifting the knocker ; and every farmer knows, to his frequent vexation, how readily cows familiarize themselves with the mechanism of gates.

Schlagintweit states that in India wild elephants

---

* Verhandlungen der k. k. zoologisch-botanischen Gesellschaft. Wien., Bd. xviii, pp. 905, 906.

build walls of sand and stones across the dry beds of rivers, in order to keep the water from flowing off during the rainy season, so as to have a sufficient supply in times of drought. The native inhabitants do the same; but whether the elephants or the Hindus were the original builders of these dams is not recorded. If such constructions imply forethought and mechanical skill on the part of man, they presume the existence of the same faculties to an equal degree in the animal.

Crows, cormorants, gulls, and other birds carry shellfish into the air and drop them on rocks, in order to break their hard covering and to eat the flesh. If the first fall is not sufficient, they carry it up still higher, and thus virtually hit it a harder blow. If a boy cracks a nut by hurling it against a stone, he makes use of the stone as a tool as truly as if he should take a stone in his hand and strike the nut with it. The former process is that employed by the birds, which are in this respect tool-using animals. There are rocks on the seacoast which have served generations of birds as stationary hammers for smashing mollusks, and are evidently regarded by them as a permanent slaughter-house.

It is well known that monkeys living near the seashore, where the ebb tide leaves the rocks covered with oysters, evince extraordinary expertness in opening these bivalves with sharp stones, just as a man would do under like circumstances. It would require only a very slight increase of intelligence for a monkey to learn to break a stone into proper shape, instead of selecting a suitable one from the shingle of the beach, and, by thus fabricating a tool, bring himself abreast, intellectually, with the flint-clipping man of the early Stone age. Indeed, it has been suggested by some scientists that man had not yet appeared upon the earth in the Miocene age, and that the chipped flints of that period are the work of semihuman pithecoid apes of superior intelligence; and there is nothing in the theory of evolution or the facts of natural history which would render such a supposition absurd. Monkeys use stones as hammers and sticks as levers, and appreciate the advantage to be derived from this the simplest of the mechanical powers. With them, as with primitive or uneducated men, this knowledge is purely empirical, a product of experience, and does not imply a perception of mathematical truths or principles any more than the taking of a short cut diagonally across a field involves a knowledge of the relation of the hypothenuse to the other two sides of a right-angled triangle. In neither case is there any question of what Prantl calls " spontaneous mathematical thinking."

Dr. Macgowan, who has resided in China since 1843, and travelled extensively in the Flowery Kingdom, states in a recent number of the North China Daily News (1893) that there exists in the mountainous and densely wooded region of Manchuria near the Great Wall a species of ape which prepares from berries two sorts of wine, one greenish and the other reddish, and preserves them in earthern jars for winter use, when the springs and rivers are frozen. The jars are also made by the apes and are fully equal in workmanship to the pottery of many savage tribes. Dr. Macgowan asserts that there is in the province of Chekiang a kind of orang-outang which shows the same skill and prudence in manufacturing and storing beverages for the time of need. It is possible that the jars of wine may have been stolen by the monkeys, although the mountains in which they live are not inhabited by human beings. According to Chinese authorities the orang-outangs of Chekiang have been observed pounding berries and other fruits in stone mortars.

Simian dexterity is greatly increased by association with human beings and by observation of their doings. The owner of a pet monkey, which annoyed him by ringing the servants' bell, tied several knots in the cord, in order to make it shorter and place it out of the animal's reach. But the crafty creature was not to be thwarted by such a clumsy device, and, climbing up on a chair, artfully untied all the knots, and then gave the bell a succession of violent jerks to signalize his triumph.

A monkey in the Zoölogical Garden at Philadelphia came into possession of a marble and a hickory nut which he tried in vain to crack with his teeth. After conferring with two other monkeys and chattering in a lively manner, he scraped away the sawdust so as to expose a space about two feet square of the zinc floor of his cage. He then climbed up on a crossbar, and from this vantage ground hurled the marble with all his force against the zinc and broke it into pieces, but found nothing edible inside. He then attempted to break the nut in the same manner, but without success. After several futile efforts he held another consultation with his companions and then handed the nut through the bars to a bystander, who cracked and returned it. The monkey then divided it into three portions, of which he gave one to each of his friends and advisers. The monkey in this case acted as a child would have done under similar circumstances, and showed a like degree and kind of reflection and ingenuous confidence. It is probable, too, that his conduct was somewhat influenced by previous study of mankind.

In the Transvaal monkeys have been found to be very serviceable in the gold mines. Captain E. Moss states that he has twenty-four monkeys thus employed and that they do the work of about seven able-bodied men, and "it is no reflection upon the human labourers to say that they do a class of work a man can not do as well as they. In many instances they lend valuable aid where a man is useless. They gather up the small pieces of quartz that would be passed unnoticed by the workmen, and pile them up in little heaps that can be easily gathered up in a shovel and thrown into the mill. They are exceedingly adept at catching the little particles, and the very things that the human eye would easily pass over never escape their sharp eyes." He then relates how the idea of thus making use of them first occurred to him, and it is a pleasure to learn that they are not forced to work, but simply do of their own free will what they see others doing.

"When I went digging gold I had two monkeys that were exceedingly interesting pets. They were constantly following me about the mines, and one day I noticed that they were busily engaged in gathering up little bits of quartz and putting them in piles. They seemed to enjoy the labour very much, and would go to the mines every morning and work there during the day. It did not take me long to learn their value as labourers, and I decided to procure more. So I immediately procured a number, and now have two dozen working daily in and about the mines. It is exceedingly interesting to watch my two pet monkeys teach the new ones how to work, and still stranger to see how the newcomers take to it. They work just as they please, sometimes going down into the mines when they have cleared up all the *débris* on the outside. They live and work together without quarrelling any more than men do. They are quite methodical in their habits, and go to work and finish up in the same manner as human beings would do under similar circumstances."

Prantl also characterizes man as the only animal familiar with the use of fire, and capable of applying it to culinary and economical purposes and to the increase of personal comfort. But this attainment is by no means common to all mankind. *Homo sapiens* inhabited the earth for ages before he discovered methods of generating this element and making it subservient to his interests. The habitual use of fire is the sign of a very considerable advancement toward civilization, and marks an important epoch in the evolution of the race. Chimpanzees, gorillas, and orang-outangs have been repeatedly seen bringing brushwood and throwing it on the camp fires which travellers have left burning, showing that they have learned by observation how to keep up a fire, although they have no means and do not understand the art of kindling it. By associating with man they soon acquire this knowledge, igniting friction matches, and often have to be watched carefully, like children, lest they should do immense mischief unwittingly as incendiaries. The same is true of ravens, which, when tamed, are fond of throwing pieces of paper and other light combustibles on the glowing coals, and seeing them flash into flame. This favourite pastime renders them exceedingly dangerous inmates of the house; and it is probably this bird that was spoken of by Pliny as *avis incendiaria*.

Ants store in their chambered hillocks certain substances which, by fermentation, produce quite a high temperature, and are put there for the sole purpose of generating heat and warming their dwellings. Some birds, as, for example, the Australian megapode, or jungle fowl, hatch their eggs by artificial heat, resulting from the decomposition of the leaves and decaying substances with which they cover them; raising large mounds that are sometimes twenty or thirty metres in circumference, and serve as incubators for successive generations of birds. Thus, while it is true that animals do not make use of fire, they are not ignorant of the properties of heat, which they turn to practical account in matters of domestic economy and household life.

It is questionable whether Prantl's statement that animals "expect an effect, but not a logical sequence, and seek a cause, but not a logical ground," can be maintained. The following incident, related by Dr. Schomburgk, director of the zoölogical garden at Adelaide, in South Australia, would seem to render such a distinction untenable. An old monkey of the genus *Macacus sinicus*, which was confined in a cage with two younger ones, flew at the keeper one day as he was supplying them with fresh water, and bit him so severely in the wrist as to injure the sinews and artery and to endanger his life. Schomburgk ordered the animal to be shot, but as an attendant approached the cage with a gun the culprit showed the greatest consternation, fled into the sleeping apartment of the cage, and could not be induced by any offers of tempting food to come out of this place of refuge. It must be added that the monkeys were perfectly accustomed to firearms, which had been frequently used for killing rats near the cage, and had never manifested the slightest fear of them. Even now the other monkeys ate their food as usual, with a conscience void of offence, and were not at all disturbed by the sight of the murderous weapon. No sooner had the man with the gun withdrawn and concealed himself than the old monkey sneaked out, and, snatching some of the food, rushed back into his asylum; but when he tried to repeat this experiment a keeper closed the sliding-door from without, and thus cut off his retreat. As the man with the gun drew near again, the poor monkey seemed quite beside himself with terror. He first tried to open the sliding-door, then ran into every nook and corner of the cage in search of some way of escape, and finally, in despair, threw himself flat on the floor and awaited his fate, which soon overtook him. The conduct of the monkey in this case can be explained only by assuming the animal to have been endowed with a moral sense and a logical faculty, implying a clear perception of right and wrong, a consciousness of guilt, a knowledge of the use of firearms, and quite a complicated process of reasoning from these premises to a perfectly correct conclusion.

The following case of quite recent occurrence proves that other animals besides monkeys are capable of reasoning deductively. Mr. Allen H. Norton, the owner of a farm at Winsted, Conn., had a dog of mixed breed, partly cocker spaniel and partly hound, which was a good hunter and had been very serviceable in catching raccoons and other small game. In the spring of 1897 the dog, now twelve years old, was getting rather feeble and had lost some of its teeth, so that Mr. Norton, thinking that he had no further use for it, since it had ceased to be useful to *him*, concluded to have it shot. For this purpose he gave it over to the tenant, who took it into the field, put his gun on the ground, and began to dig a grave for the faithful animal, which lay beside the weapon intended soon to end its life, and watched the hole as it gradually grew deeper. When the work was nearly finished the dog suddenly sprang to its feet and ran away in great haste. The man tried to call it back, but for the first time on record it refused to obey, and rushing to the bank of the river, swam to the opposite side, disappeared in the woods, and never returned. This instance is even more remarkable than that of Dr. Schomburgk's monkey, since the acute exercise of the logical faculty was not stimulated by the prickings of conscience.

Perhaps the most human of anthropoid apes, as regards intelligence, is a species of chimpanzee called the soko, first discovered by Livingstone, and most fully described by him in his Last Journals. The teeth of these creatures, he says, "are slightly human, but their canines show the beast by their large development. The hands, or rather the fingers, are like those of the natives. They live in communities consisting of about a dozen individuals, and are strictly monogamous in their conjugal relations, and vegetarian, or rather frugivorous, in their diet, their favourite food being bananas." The aborigines, the Manyuema, are, on the contrary, cannibals, and are described by Livingstone as "the lowest of the low." One of them, who had killed a woman, offered his grandmother to be killed in expiation of his offence, and this vicarious punishment was accepted as satisfactory. Even the sokos have a higher and more correct conception of justice than this; at least they do not make the innocent atone for the crimes of the guilty. If a soko "tries to seize the female of another, he is caught on the ground, and all unite in boxing and biting the offender." "Numbers of them come down in the forest within a hundred yards of our camp, and would be unknown but for giving tongue like foxhounds. This is their nearest approach to speech. A man hoeing was stalked by a soko and seized. He roared out, but the soko giggled and grinned, and left him, as if he had done it in play." It is evident that these animals have some sense of humour and appreciate a practical joke. They are inoffensive and unaggressive, but fearless and energetic in self-defence. They never molest women or unarmed men, but if any one approaches them with a spear they rush upon him and wrest the weapon from his hands. If struck with a dart or an arrow, they pull it out, and stanch the blood by stuffing leaves into the wound. The natives recognise their harmless and human character, and say, "Soko is a man, and nothing bad in him."

Sometimes they kidnap a child and take it up into a tree, but they never hurt it, and are ready to exchange it at any time for a bunch of bananas. Perhaps the robbery is for the sake of the ransom. When roaming through the forest, the female usually carries her infant in her arms; but in crossing a glade or other open ground, where they would be more exposed to danger, the father takes the child, and returns it to the mother as soon as they enter the wood again. They are extremely fond of assembling in a remote part of the forest and drumming on hollow trees and other resonant objects, accompanying this fearful din with loud yells, like sopranos and tenors of strong pulmonary powers trying to outshriek the clash and clang of a Wagnerian orchestra. This deafening noise does not differ greatly from " the natives' embryotic music," and is quite as harmonious and pleasant to the ear as much of the music of the Chinese and other Oriental peoples.

Livingstone had a young female soko, which, after having been petted for some time, was " quite like a spoiled child." She enjoyed shaking hands, and took as much pleasure in this tiresome manual ceremony as any American citizen who honours the President of the United States by calling on him at the White House. She liked to be carried about, and would beg people to take her in their arms. If they refused, she seemed greatly aggrieved, and would make a wry face, as if about to burst into tears, and wring her hands, apparently in severe distress of mind. She learned to eat whatever was set before her, drew grass and leaves around her for a bed, and covered herself with a mat when she went to sleep. She could untie a knot with her fingers and thumbs "in quite a systematic way," "looked daggers" at any one who interfered with her doings, and resented every attempt to touch what she regarded as her personal property.

Indeed, the idea of personal property, in distinction from communal property—such, for example, as the provisions stored by ants for winter—is quite as strongly developed in many of the higher species of animals as in some of the lower races of men.

CHAPTER VIII.

SPEECH AS A BARRIER BETWEEN MAN AND BEAST.

Max Müller's theory. Hobbes's pun. Roots as ultimate facts. Sanskrit illustrations of their formation. Example of "quack" as a prolific root. Dr. Hun's specimen of child language. Horatio Hale's theory of the origin of tribal dialects. Metamorphoses of organic life and transformations of the roots of speech. A philological ultimatum. General concepts improperly denied to animals. Horne Tooke's absurd statement. Ability of animals to count and to classify objects. Words not the only symbols of thought. Whitney's doctrine of language as a social institution. Pantomimic expression in man and animals. Jäger's distinction between emotional language and the language of thought. Worthlessness of speculation without careful observation. Aphasia. Animals learn to understand human speech. "Nursery philology." Language evolved out of roots as vital organisms out of protoplasm. Clifford's poetic description of atoms. Noiré's fantastic synergastic theory. The *Homo alalus* as a social being. Speech not a supernatural endowment. Bow-wow, pooh-pooh; and yo-he-ho theories. Psophos *vs.* phônê. Animal utterances not mere unconceptional noises. Landois on animal voices. No break in the evolution of expression. Weir and Janet on the vocal organs of ants. Philology in the menagerie. Garner's studies of simian speech. Successful use of the phonograph. Roots and concepts in the language of animals. Hypothetical language of the "missing link." Investigations of animal speech by Wenzel, Radeau, Jules Richard, and others. Remarkable parrots. Superior advantages possessed by Garner for prosecuting his researches. His failure to accomplish

what he expected to do in Africa. Dybowski's account of it.
His own report of the expedition confirmatory of Dybowski's
criticism.

MAX MÜLLER, after admitting "the extraordinary accounts of the intellect, the understanding, the caution, the judgment, the sagacity, acuteness, cleverness, genius, or even social virtues of animals," intrenches himself behind the "one palpable fact, namely, that, whatever animals do or do not do, *no animal has ever spoken.*" This assertion is not strictly true. Parrots and ravens utter articulate sounds as distinctly as the average cockney, and in most cases make quite as intelligent and edifying use of them for the expression of ideas.

That no animal has ever made a natural and habitual use of articulate speech for the communication of its thoughts and feelings is a truism which it would seem superfluous to emphasize or italicize. Equally irrelevant to the point at issue is the statement that "in every book on logic language is quoted as the specific difference between man and other beings." It is not by the definitions of logicians that questions of this kind are to be decided. The Greeks called beasts speechless creatures (τὰ ἄλογα) just as they called foreigners tongueless (ἄγλωττοι), meaning thereby persons whose language was unintelligible to them; and the epithet was no more appropriate in the former case than in the latter. It was for the same reason that the Roman poet Ovid, when banished to the Pontus, characterized himself as a barbarian, because his language was not understood by the inhabitants of that country—*barbarus hic ego sum, quia non intelligor ulli.* But such expressions must not be taken too literally.

Hobbes makes speaking the test of rationality—*homo animal rationale, quia orationale*—and assumes both powers to be the exclusive property of man; but his pithy statement is a quibble in fact as well as in form, and much better as a pun than as a psychological proposition. "Language is our Rubicon," says Max Müller, "and no brute will dare to cross it." Why not? Because, if he does, our definitions will transform him from a brute into a man. "In a series of forms graduating from some apelike creature to man," Max Müller maintains that the point where the animal ceases and the man begins can be determined with absolute precision, since "it would be coincident with the beginning of the radical period of language, with the first formation of a general idea embodied in the only form in which we find it embodied, namely, in the roots of our language."

In reply to the statement that "both man and monkey are born without language," Müller asks "why a man always learns to speak, a monkey never." This query, if it is to be regarded as anything more than a bit of banter, implies a gross misconception of the theory of evolution, as though it involved the development of an individual monkey into an individual man. One might as well deny the descent of the dog from the wolf because a dog always learns to bark, a wolf never. In the course of ages, and as the result of long processes of evolution and transformation, monkeys have learned to speak, but when they have acquired this faculty we call them men.

Max Müller stops at roots or "phonetic cells" as "ultimate facts in the analysis of language," and virtually says to the philologist, "Thus far shalt thou go, and no farther, and here shall thy researches be stayed." "The scholar," he declares, "begins and ends with these phonetic types; or, if he ignores them, and traces words back to the cries of animals or to the interjections of men, he does so at his peril. The philosopher goes beyond, and he discovers in the line which separates rational from emotional language, conceptual from intuitional knowledge—in the roots of language he discovers the true barrier between Man and Beast."

The philologist, who recognises in the roots of language the Ultima Thule beyond which he dare not push his investigations, confesses thereby his incompetency to solve the problem of the origin of language, and must resign this field of inquiry to the zoöpsychologist, who, freeing himself from the trammels and illusions of metaphysics, seeks to find a firm basis for his science in the strict and systematic study of facts. Imagine the folly of the physiologist who should say to his fellow-scientists: "In your researches you must begin and end with cells. If, in studying organic structures, you go back of cells and endeavour to discover the laws underlying their origin, you do so at your peril. Beware of the dangerous seductions of cytoblast and cytogenesis and the treacherous quagmires of protoplasm."

Nevertheless, this attitude of mind is natural enough to the philologist, who is so absorbed in the laws which govern the transmutations of words that he comes to regard these metamorphoses as finalities, and never goes behind and beyond them. We must look, therefore, not to comparative philology, but to comparative psychology, for the discovery of the origin of language. Philology has to do with the growth and development of speech out of roots, which are assumed to be ultimate and unanalyzable elements, like the purely hypothetical particles which the physicist calls atoms; but as to the nature and genesis of roots themselves the philologist of to-day is as puzzled and perplexed as was the old Vedic poet when, in the presence of the universe and its mysterious generation, he could only utter the pathetic and helpless cry, "Who indeed knows, who can declare, whence it sprang, whence this evolution?"

Doubtless the emotional stage precedes the intellectual or rational stage in the growth of language, but the former mode of expression does not cease when the latter begins, nor is it possible to draw a fixed and fast line of demarcation between them. *Pâ* and *mâ* are the roots of *pâtri* and *mâtri*, and mean in Sanskrit to protect and to form, indicating the function of the father as the defender, and of the mother as the moulder, of children. But how did they come to have these significations? Surely the infant who first used these expressions—and they are universally recognised as belonging to the vocabulary of babes—did not associate with them the ideas which philologists now discover, and which grammarians and etymologists at a very early period put into them. How arbitrary these inferences are is evident from the variety of interpretations of which such words are susceptible. Thus *mâ* means also to measure; hence the moon, as the measurer of time, was called *mâtri;* and from this point of view the term for mother was explained as referring to her office as the head of the household, who kept the keys of closet and pantry, and meted out to the servants and other members of the family the things necessary for them. It is furthermore a suspicious circumstance touching the habits of the Indo-Aryan's progenitors that *pâ* means to drink, and *pâtṛi* signifies a drinker; and for aught we know the verbal coincidence may not be accidental. As regards *mâ*, it means also bleating as a goat, and occurs in this sense in the Rig-Veda; and it is probable that in this onomatopoetic expression we come nearer to the real origin of the word for mother.

There is a vast deal of vague speculation and untenable assertion concerning the origin and formation of roots in language. In Sanskrit, for example, there are three radical words *gar*, meaning respectively to swallow, to make a noise, and to wake. It is conceivable, says Max Müller, that the first two of these roots may have been originally one and the same, and that *gar*, from meaning to swallow, may have come to mean the indistinct and disagreeable noise which often attends deglutition, and which in speaking is called swallowing letters or words. Yet the third root, he adds, can hardly be traced back to the same source, but has the right to be treated as a legitimate and independent companion of the other roots. From this example he deduces the general principle that if roots have the same form, but a different meaning, they are to be regarded as originally different, notwithstanding their outward resemblance. He then passes from etymology to embryology, and reasons from analogy that "if two germs, though apparently alike, grow, under all circumstances, the one always into an ape and never beyond, the other always into a man and never below, then the two germs, though indistinguishable at first, and though following for a time the same line of embryonic development, are different from the beginning, whatever their beginning may have been."

In this statement he begs the whole question at issue; and the philological illustration which he brings to bear upon an anthropological theory for the purpose of refuting it is itself exceedingly questionable, since nothing is easier or would be more natural than to derive *gar*, to wake, from *gar*, to make a noise; so that all three roots not only may have had, but probably did have, a common origin. In no case can it be positively affirmed that roots of the same form are not of the same origin, however widely they may differ from one another in signification.

One of Darwin's grandchildren, as Mr. Romanes states, called a duck "quack," and by a special and easily intelligible association called water also "quack." The same term was afterward extended to all fowls and winged creatures and to all fluids. A French sou and an American dollar were called "quack" on account of the eagle stamped upon them, and the same name was then given to all coins. Thus "quack" came to mean bird, fly, angel, wine, pond, river, shilling, medal, etc., and it is easy to trace every step of the process by which it acquired these various significations.

According to Max Müller's reasoning, "quack" in the sense of duck or bird must have a radically different origin from "quack" in the sense of pond or shilling. But how do we know that all roots having the same form, but different meanings, may not have originated in this manner? Because we can no longer trace a word through all phases of its development and metamorphosis is no proof that the development and metamorphosis never took place. The evolution of the word "quack" in the vocabulary of the aforesaid child shows furthermore that a purely onomatopoetic root is not always sterile, but may be prodigiously and puzzlingly prolific, germinating in the mind of the primitive man, and springing up and bearing fruit fifty or a hundredfold.

When we speak of a train of cars as "telescoped," this use of the word has nothing in common with its primary and etymological meaning, and can be understood only by a knowledge of the construction of a telescope out of concentric tubes sliding into each other. Again, the telescopic chimney of a war vessel is not a point of far-seeing observation, as the composition of the qualifying word would imply, but a chimney which may be shoved together endwise, and thus put out of reach of the enemy's shot.

Dr. Hun records in The Monthly Journal of Psychological Medicine (1868) the case of a girl who invented a language of her own, and taught it to her younger brother. *Papa* and *mamma* used separately meant father and mother; but when linked together in the compound *papa-mamma* they meant church, prayer book, praying, and other acts of religious worship, because the child saw her parents going to church together. *Gar odo* meant "Send for the horse," and also paper and pencil, because the order for the horse was often written. *Bau* signified soldier and bishop, because both seemed to be more gorgeously dressed than other persons. Here the clothes made the man, and furnished the sole basis of his classification. It needed only the simplest and most superficial point of association in order to attach the most diverse significations to the same word.

To the objection that these examples are mere childish whimseys, and that languages never originate and grow up in this manner, it may be replied that such an assertion assumes the very point to be proved. Mr. Horatio Hale maintains that the aboriginal tongues of South America and South Africa were produced in precisely this way. He thinks, too, that the numerous tribal dialects west of the Rocky Mountains had their origin in the isolation of orphaned children, and that such a result is possible, and indeed inevitable, wherever the climate and other external conditions are favourable to the survival of small children bereft of their parents and separated from their kinsmen.

Again, Max Müller observes, in explanation of the manner in which roots were formed, that, "after a long struggle, the uncertain phonetic imitations of special impressions became the definite phonetic representations of general concepts." Thus "there must have been many imitations of the falling of stones, trees, leaves, rivers, rain, and hail, but in the end they were all combined in the simple root *pat*, expressive of quick movement, whether in falling, flying, or running. By giving up all that could remind the hearer of any special sound of rushing objects, the root *pat* became fitted as a sign of the general concept of quick movement." There was a great number of "imitative sounds of falling, out of which *pat* was selected, or out of which *pat*, by a higher degree of fitness, struggled into life and fixity." So, too, the prolific root *mar*, to grind or to break, " must be looked upon as tuned down from innumerable imitations of the sounds of breaking, crushing, crunching, crashing, smashing, mashing, cracking, creaking, rattling and clattering, mauling and marring, till at last, after removing all that seemed too special, there remained the smooth and manageable Aryan root of *mar*."

Now, pray, when did this remarkable evolution, which implies the close and continuous exercise of rare powers of comparison and abstraction and the perfect maturity of the intellectual faculties, take place? "Language," we are informed, "presupposes the formation of concepts," and "all such concepts are embodied in roots." The formation of these concepts, then, must have preceded, logically and chronologically, the formation of the roots in which they are embodied, and must therefore have been effected without the aid of language, which was subsequently evolved or elaborated out of these roots. What becomes, then, of the assertion that it is impossible to think or to generalize without language, since language itself originated in a long and laborious process of thought and generalization?

The manner in which the word "quack," in the case already cited, gradually acquired its widely different meanings is perfectly intelligible. Suppose, now, that the child, after having grown to manhood, retained, as the result of isolation, the use of the word "quack" in its diverse significations, and taught and transmitted it to his posterity, so that it became incorporated in the language of his race. In a few generations, especially among a rude people, the origin of the word would be forgotten, and it would be difficult to imagine how it came to acquire such a variety of meanings, and to stand for so many objects having apparently no connection with one another. In due time the philologist would come with his *apparatus criticus*, subject the word to a strictly scientific analysis, apply all the approved tests, and, after great expenditure of etymological erudition and conjectural ingenuity, would discover half a dozen wholly independent roots of "quack" which could not be traced to one and the same source.

No one knows how often, in the formative period of language, it may have happened that the growth of a word and the multiplication of its meanings may have been obscured and rendered incomprehensible because the intermediate stages of its development were forgotten, and the connecting links that made the transition easy and natural were lost. In the instance just cited we have also an example of a fruitful onomatopoetic root. Indeed, in our own tongue, "quack," the mere imitation of an animal cry, has given rise to a variety of words and conceptions, such as quack, quacksalver, quackery, which are as remote in their relations to the web-footed fowl as is the man who "plays at ducks and drakes" with his money, and ends his career as a "lame duck."

Nothing could be more abrupt or incredible, to take an illustration from Nature, than the metamorphoses of the Lepidoptera, the same individual undergoing the most marvellous changes from caterpillar into chrysalis, and again into butterfly. Here the transformations are so great that, if we saw merely the result, we should never suspect the nature of the process. Creatures that for a long time were supposed to be entirely distinct, and were classified as belonging not only to different genera, but even to different orders of animals, are now known to be the same individual in different phases or stages of its development. Thus, as we are told by an eminent authority on crustacea, "the *Zoëa*, the *Megalops*, and the *Carcinus Mœnas*, or shore crab, are but the baby, the child, and the adult forms of a single individual."

The *Amphicyon* is an animal which may have been the common ancestor of the dog and the bear, although more closely allied to the former than to the latter. The *Hyænarctos*, on the contrary, possesses more ursine than canine characteristics, but, by change of environment and under stress of circumstances, might have branched off in either direction. The *Archæopteryx Lithographica* was a sort of griffin, from which both birds and reptiles may have descended. The *Phenacodus Primævus* may have been the progenitor of hoofed animals or clawed animals, and needed only slight modifications in order to ramify into either class of quadrupeds. Examples of this sort abound among fossil creatures.

What is here shown to be true of living organisms is still more probable of roots of speech; and the naturalist might, with at least equal cogency and validity, argue analogically from the identity of these so exceedingly diverse crustacea, or from the common origin of man and ape, that roots like *dâ* and *gar*, however much they may differ in meaning, are really traceable to one and the same source.

"Show me only one root in the language of animals," says Max Müller, "such as *ak*, to be sharp and quick, and from it two derivatives, as *asva*, the quick one—the horse—and *acutus*, sharp or quick-witted; nay, show me one animal that has the power of forming roots, that can put one and two together, and realize the simplest dual concept; show me one animal that can think and say 'two,' and I should say that, so far as language is concerned, we can not oppose Mr. Darwin's argument, and that man has, or at least may have been, developed from some lower animal."

Nothing could be more absurd than this sort of philological ultimatum, since, according to the theory of evolution, the language of animals has not yet reached this stage of development; for it would then become articulate speech, and be no longer the language of animals, but the language of man. But this is surely no evidence or indication that one may not grow out of the other; on the contrary, it rather suggests the possibility of such growth and development.

We can not be certain, however, that animals may not have general concepts. When a dog, in eager pursuit of some object, yelps *ak-ak*, how do we know that this sharp utterance, which expresses the strong and impatient desire of the dog to overtake the object, may not stand in the canine mind for the general concept of quickness? It is used in pursuing all animals and inanimate things, bird, hare, squirrel, stick, or stone, and cannot therefore denote any single one of them, but must have a general signification. For aught we know, the language of animals may be made up of undeveloped roots vaguely expressive of general concepts, or may even contain derivative sounds. The bark of a dog after bringing a stick or a stone to its master and requesting him to throw it again is slightly different from the sharp yelp uttered in pursuing it; and it is impossible to know whether these sounds may not stand to each other in the relation of the radical to its derivative.

Darwin asserts that "the dog, since being domesticated, has learned to bark in at least five or six distinct tones, namely: the bark of eagerness, as in the chase; that of anger, as well as growling; the yelp, or howl of despair, when shut up; the baying at night; the bark of joy, when starting on a walk with his master; and the very distinct one of demand or supplication, as when wishing for a door or window to be opened." This variety of tones, expressing different desires and emotions in an animal that in its wild state could not bark at all, marks a very considerable advance in the power of vocal utterance as the result of association with man.

Max Müller has recently come to the conclusion that roots originated in cries uttered by men in performing certain actions, such as digging, cutting, lifting, or pounding. This so-called *clamor concomitans*, or sound attending the action, became by association a *clamor significans*, or sound signifying the action. This explanation of the genesis of roots is doubtless, to a certain extent, correct, but comes perilously near to the "bow-wow" and "pooh-pooh" theories which he formerly rejected with ridicule and ineffable scorn. It would be hard, however, to find a finer combination of concomitant and significant clamour than the deep bay of a pack of hounds.

In one of his lectures Müller quotes, as "an excellent answer to the interjectional theory," the following observations of Horne Tooke in the Diversions of Purley: "The dominion of speech is erected upon the downfall of interjections. Without the artful contrivance of language, mankind would have had nothing but interjections with which to communicate orally any of their feelings. The neighing of a horse, the lowing of a cow, the barking of a dog, the purring of a cat, sneezing, coughing, groaning, shrieking, and every other involuntary convulsion with oral sound have almost as good a title to be called parts of speech as interjections have. Voluntary interjections are employed only when the suddenness and vehemence of some affection or passion return men to their natural state, and make them for a moment forget the use of speech; or when, from some circumstance, the shortness of time will not permit them to exercise it."

This passage really confirms in the strongest manner the theory which it is cited in order to refute. The dominion of every improved implement is founded upon the downfall of an inferior implement. Thus the steel plough has superseded the pointed piece of wood with which the primitive husbandman scratched the surface of the earth; the matchlock has supplanted the crossbow, the Remington rifle the rude musket, and the steam car the old stagecoach. Everywhere in the progress of human invention the better instrument takes the place of the poorer one and robs it of its supremacy. The evolution of language furnishes no exception to this universal law. It is a means of communicating ideas and emotions from one person to another, and the more clearly, concisely, and forcibly it performs this function the more perfect it is as an instrument. To speak of the grammatically complicated, and therefore practically clumsy, Sanskrit as superior to the simple and handy English, and to characterize the latter as the result of degeneration and decay, is an abuse of terms involving an utter misconception of the purpose for which language exists. Sanskrit may be more interesting philologically than English, just as the five-toed *Eohippus* and the three-toed *Hipparion* may be more interesting anatomically than the horse; but no one would deny that the modern quadruped combines in a greater degree simplicity of structure with efficiency of function, and is therefore, as an animal, superior to its ancient prototypes.

The very fact that, as Horne Tooke observes, men return to their natural state in the use of interjections and exclamations well-nigh proves that these are the raw material, or linguistic protoplasm, out of which articulate or organic speech was evolved. But to compare a cough and a sneeze to an interjection, or to put them in the same category with the neigh of a horse, the bark of a dog, or the purr of a cat, shows a strange lack of discrimination between purely physical and involuntary convulsions and vocal sounds intended to express emotions of the mind. A cough or a sneeze may be more or less successfully imitated, like a stage laugh, and thus become the sign and suggestion of an idea; but a genuine cough or sneeze is a violent expulsion of the air through the throat or nose in consequence of local irritation beyond a man's control, and has, therefore, no oral or intellectual element in it.

As regards the ability of animals to "think and say 'two,'" it has been proved conclusively that the magpie and some other birds, even in their wild state, can count at least four, and this fact is recognised and utilized by fowlers; but if it be true that it is impossible to form the concept "four" without the aid of language, it follows that the magpie must be able to say "four" in a language of its own. To deny this conclusion because we do not understand "margot" (as the magpie language might be called) would be to set up our own ignorance as a standard by which to test the magpie's intellectual capacity, and thus fall into the fallacy of *argumentum ab ignorantia facti*. This knowledge of numeration can be greatly extended by instruction. A chimpanzee in the London Zoölogical Gardens, says Mr. Romanes, has been taught to count five. Ask her for four, three, two, or five straws in any order of succession, and she will give the exact number required. She understands not only the names of these numerals, but also other words and phrases, just as a child does before learning to speak.

All classification rests upon the power of generalization, and this faculty belongs to the lower animals as well as to men. As has been remarked by an acute observer: "Dogs can distinguish strangers and acquaintances, well-dressed persons from persons in rags, the canine species from all other species. They can not carry their classification far, not from want of memory and intelligence, but from want of a well-defined language and printed books." The dullest dog has a lively perception of the difference between canine and feline. No matter how much the particular dog may vary from other individuals of the species,

> As hounds and greyhounds, mongrels, spaniels, curs,
> Shoughs, water-rugs, and demi-wolves,

he is never confounded with the cat, but is at once recognised as canine. The dog not only thinks of these so diverse creatures as belonging to the same class, but is also conscious of belonging to it himself. Man's intellectual superiority consists in possessing a greater number of these concepts, and in being able to compare and combine them in reasoning processes with greater accuracy and facility, than the beast, although there are tribes of men in which this superiority is so slight as to be scarcely perceptible.

Thus, for example, the aborigines of Australia have

no words for the expression of abstract ideas or general conceptions, not even collective names for animals and plants, indicating a lack of the faculty of generalization. Their ability to discriminate between animals of the same species is far greater than the poverty of their vocabulary would imply. They do not mistake a wild duck for a wild goose, and yet they call them both "*monarum.*" Every poisonous serpent is a "*wonge*" and every kind of turtle a "*miaro.*" White is "*bambar*" and black is "*ngurue,*" but red, green, blue, and yellow are lumped together as "*beiar*"; but it would be incorrect to infer from this want of special designations that they do not distinguish between these four colours with the eye. The development of language has not kept pace with the cultivation of the organs of sense. Some Australian tribes can count only three, and none of them more than five; "*garro*" is one, "*boö*" two, "*koromde*" three, "*wogaro*" four, and "*boö koromde*" five; more than five is "*meian,*" many. Even the natives who have learned a little English are not able to count beyond six with any degree of certainty, although they make use of the English numerals.*

Exclamations, according to Max Müller, "are as little to be called words as the expressive gestures which usually accompany these exclamations." No one asserts that they are words in the strict sense of the term; all that is claimed for them is that they express thoughts and feelings or reveal states of the mind, and may be regarded as language. This he admits when he adds, "In fact, interjections, together with gestures, the

---

* Im australischen Busch, von Richard Semon, Leipzig, 1896, p. 240.

movements of the muscles of the mouth and the eye, would be quite sufficient for all purposes which language answers with the majority of mankind." But as such exclamations and gesticulations are not words and do not constitute language, the majority of mankind are destitute of thought, since we are assured that "language and thought are inseparable," and that "there is no thought without words, as little as there are words without thought."

Prof. Mansel is nearer the truth when he says, "As a matter of necessity, men must think by symbols; as a matter of fact, they do think by language." But although words are the most convenient and most perfect symbols of thought, they are by no means the only ones. A man can count three by holding up three fingers, or by touching three objects, or by laying down three sticks, as the Veddahs do in bartering, without the aid of articulate speech. A dog can do the same by barking three times. It is not true that "language begins where interjections end." Articulate speech begins where pantomimic expression, emphasized by mere hooting and hallooing, ends; but both are instruments of thought and symbols for the representation and communication of ideas.

"Speech," as Prof. Whitney has justly observed, "is not a personal possession, but a social institution. What we may severally choose to say is not language until it is accepted and employed by our fellows. The whole development of speech is wrought out by the community. That is a word, and only that, which is understood in a community. Their mutual understanding is the tie that connects it with the idea. It is a sign which each one has acquired from without, from the usage of others." Goethe, in his epigram Etymologie, expresses the same thought:

> So wird erst nach und nach die Sprache festgerammelt,
> Und was ein Volk zusammen sich gestammelt,
> Muss ewiges Gesetz für Herz und Seele sein.

"Man," says Wilhelm von Humboldt, "understands himself fully only by testing the intelligibility of his words on others. The objectivity is increased when the word which he has formed is echoed back to him from the mouth of another. At the same time, it is not thereby robbed in the least of its subjective character, since man feels himself always one with man." What is felt and expressed by the individual must be refelt and re-expressed by the mass and stamped with its indorsement before it is accepted as speech.

One who is deaf and dumb from his birth learns to give digital instead of lingual expression to his thoughts, and it has been observed that such a person in the act of thinking almost unconsciously moves his fingers, as though thought and digital action (as a substitute for articulation) were necessarily and inseparably connected, thus proving that, while speech is a natural instrument for the expression of thought, other purely conventional methods may be substituted for it, and through habit may become quite as strongly associated with the thinking processes.

Among savage tribes, and even among a people so highly civilized as the Arabs, signs and gestures play a very important part in the expression of thought, and the Neapolitan's love of pantomime and skill in the use of it are well known. Of the Veddahs of Ceylon Sir James Emerson Tennent says, "So degraded are some of these wretched outcasts that it has appeared doubtful in certain cases whether they have any language whatever"; and Mr. G. R. Mercer, who, by a long residence in their country, acquired an intimate knowledge of their habits, affirms that "even their communications with one another are made by signs, grimaces, and guttural sounds which bear little or no resemblance to distinct words or systematized language." It is not correct, from an anthropological point of view, to characterize the Veddahs as "degraded." They are simply primitive and undeveloped. They are the remains of the aborigines of Ceylon; and the few articulate words they utter they have learned, parrotlike, from the Singhalese, who invaded and conquered the country, and now constitute its chief population.

Lord Monboddo's seemingly absurd and much-ridiculed theory that language was formed by an assembly of learned men convened for that purpose is right so far as it affirms the conventional and communal character of articulate speech and written language; and this is doubtless all that the laird meant to imply by his rather bullish statement. He did not intend to assert that language was framed, like a political platform, by a body of men come together expressly for that object, but that it was gradually developed in consequence of their coming together as individuals, families, and communities, and endeavouring to understand one another by means of gestures and exclamations and onomatopoetic sounds. It was also the most intelligent men of their time; those who were endowed with the greatest amount of wisdom, the quickest wits, and the readiest faculty of invention; in short, the foremost men of primitive life, who contributed most to this result. Then, as now, the progress of the race was due to the impetus imparted to it by the best brains, and was less the effect of happy chance than we are fain to imagine.

Articulate speech is an immense help to the intellectual processes of induction and deduction, abstraction and generalization, but by no means essential to these mental operations. As Dr. Paul Carus observes, "The act of naming is an enormous economy of mental activity"; but it is not absolutely necessary to this kind of activity.

The fox must have an abstract idea of danger apart from any concrete form or embodiment of it; otherwise he would not be constantly on the alert, anticipating peril when it is not present. Flourens asserts, "It is a fact that beasts do not form general ideas, and it is another fact that man does form them"; he then adds: "The study of mind by mind is that which puts the final stamp upon the profound difference separating beast from man. Intelligence in beasts does not study intelligence." Buffon caps the climax of this sort of dogmatism by declaring that in animals "*c'est le corps qui parle au corps.*" A body talking to another body without the mediation of mental faculties would be a phenomenon worth seeing.

Pantomime is the natural language of man and the lower animals, and is intelligible without previous study. In this respect it differs from articulate speech, which is mainly conventional in its character. A word has the meaning which common consent has tacitly attributed to it, and which usage has sanctioned. It is not necessary, however, that any two persons should agree beforehand as to the signification of mimetic movements in order to be able to communicate their ideas in this manner. Two deaf-mutes, or savages of alien tribes, on meeting for the first time, have no definite stock of signs with which to converse, but create them as they go along. If one sign fails to express the thought clearly, they try another. If A wishes to convey to C the drift of a previous conversation with B, he will do so by means of signs many of which differ from those used in conversing with B. He will constantly invent new and more expressive signs, and thereby convey his meaning more fully and distinctly than in his first conversation. This natural sign language may be enlarged and perfected, as it is in institutes for deaf-mutes, by the introduction of conventional elements, and thus an extended mimetic system for the communication of thought may be developed.

The dog expresses thoughts and emotions by wagging his tail, to quite as good purpose as many persons do by wagging their tongues. We impart our wishes to animals almost exclusively by gestures, until they learn to understand our words, which then alone suffice, so that the pantomime is no longer necessary except for sake of emphasis in case they refuse to obey. Animals also, in communicating their desires to us, make use of signs accompanied by all sorts of vocal utterances, which through association have become intelligible.

Among insects, especially ants and bees, the language of gesture is highly developed. Owing to the smallness of these creatures, it is difficult to observe them in their conversational intercourse, and their remoteness from us in structure and organization renders it still more difficult for us to identify ourselves with them through sympathy, and to get a clear conception of their states of mind. We are fully justified, however, in inferring from their conduct that they communicate their ideas to one another with rapidity, precision, and intelligibleness. "If psychologists of to-day," remarks Prof. Wundt, "overlooking all that an animal can express through gestures and sounds, limit the possession of language to mankind, such a conclusion is scarcely less absurd than that of many philosophers of antiquity who regarded the languages of barbarous nations as animal cries."

This observation is perfectly true, but not new, inasmuch as it was made more than fourteen centuries ago by the Neoplatonist Porphyrius in his treatise on abstinence from animal food ($\pi\epsilon\rho\grave{\iota}\ \dot{\alpha}\pi o\chi\hat{\eta}s\ \check{\epsilon}\mu\psi\acute{v}\chi\omega\nu$). After stating that the different tones used by animals show that they have a language for the expression of different sentiments, such as anger, fear, and affection, he adds: "To deny animals language because it is unintelligible to us would be as absurd as for the crows to maintain that their croaking is the only rational speech, and that we are devoid of reason because we do not understand it; or for the inhabitants of Attica to claim that theirs is the only language, and that all who do not speak it are devoid of reason. Nevertheless, an inhabitant of Attica could as easily understand the language of crows as those of Persians and Syrians." Foreign tongues, to those who hear them for the first time, are hardly more intelligible than the inarticulate sounds uttered by animals. The Emperor Julian compared the speech of the Germans to the caw of ravens, and to the Athenians the conversation of Thracians and Scythians sounded like the chatter of cranes.

Prof. Jaeger's assertion that animals have merely emotional language (*Gefühlssprache*) in distinction from the language of thought (*Gedankensprache*) is psychologically untenable. In all operations of the mind, thoughts and feelings are inextricably interblended, and it is impossible to draw a line of demarcation between them. There is no language of emotion as opposed to or essentially distinct from language of thought. Emotion is only thought under tension, thought strongly emphasized and impelled by desire. Every cry or exclamation presupposes an idea or intellectual conception, without which the emotion would never arise; and it is hardly possible to determine where the one begins and the other ends.

To what an extent animals are at the mercy of metaphysicians is illustrated by the following passage from a treatise by Prof. Green: "There is no reason to suppose, because the burnt dog shuns the fire, that it perceives any relation between it and the pain of being burnt. . . . The dog's conduct may be accounted for by the simple sequence of an imagination upon a visual sensation, resembling ones which actual pain has previously followed. . . . Till dogs can talk, what data have we on which to found another explanation?" We have precisely the same data in the case of the burnt dog as in the case of the burnt child who shuns the fire; and we are justified in reasoning from analogy that the conduct of the dog is due to the same perception of cause and effect as that of the child. "The simple sequence of an imagination upon a visual sensation, resembling ones which actual pain has previously followed," means, when translated from metaphysical jargon into plain English, that, when a dog sees a flame, its resemblance to another flame which burned him leads him to avoid it, lest this one should also burn him. The misfortune of dogs in not being endowed with articulate speech is greatly aggravated if it renders them liable to have such elaborate philosophy as this mouthed over them.

The phenomenon of aphasia furnishes additional evidence that the faculty of speech is not essential to the exercise of thought or to the power of reasoning. Aphasia, or speechlessness, as has been shown by Bouilland, Broca, and other pathologists, is the result of a disease or lesion of the third frontal convolution of the left hemisphere of the brain. Any injury of this part produces a partial or complete loss of articulate speech without disturbing or diminishing in the least the action of the intellectual faculties. The vocal organs and all the mechanism of articulation remain intact, and the ability to think logically and consecutively is unimpaired. There is no paralysis of the muscular apparatus necessary to the enunciation of words, and no derangement of the mental operations so far as the formation and orderly sequence of conceptions are concerned; only the power of correct verbal expression is gone. Max Müller speaks with contempt of "a fold of the brain"; but here we have an instance in which articulate speech is dependent upon the full development and the healthy action of a mere fold of the brain, which, if his own theory be true, is the Rubicon separating man from the brute.

The aphasiac can express his thoughts and feelings by facial movements, gesticulations, and guttural noises, but is unable to articulate words correctly. He thus reverts to the condition of mankind prior to the development of the speech-producing cerebral convolution plus the knowledge and mental capacity acquired since that time. Finkelnburg reports the extreme case of a woman whose memory for things and persons was normal, and in whose general conduct nothing anomalous was observable, but who had lost entirely the use of speech, and could understand neither spoken nor written words. She was a pious Catholic, but never made the sign of the cross of her own accord or when told to do so, yet readily imitated others when she saw them do it. She was in the hospital three months, but never learned that the ringing of the bell was the signal for dinner. Symbols even of the most general character had for her no significance; her understanding was confined strictly and directly to things, and her consciousness seems to have sunk to the level of a rather dull anthropoid.

In apes, cretins, and many microcephalous persons, the convolution of the brain on which the power of articulate speech depends is rudimentary. Human and simian brains are constructed on precisely the same plan, and differ only in the development and consequent arrangement of the convolutions. "In man," says Prof. Vogt, "the third frontal convolution is extraordinarily developed and covers the insula, while the transverse central convolutions are much less prominent; in the ape, on the contrary, the third frontal convolution is but slightly developed, while the central transverse convolutions are very large, descending quite to the edge of the hemisphere and giving to the fissure of Sylvius the form of a V."

The difference is one of degree, and not of kind, resulting from the higher evolution of the same type. Max Müller admits it to be possible and intelligible that

"that most wonderful of organs, the eye, has been developed out of a pigmentary spot, and the ear out of a particularly sore place in the skin—that, in fact, an animal without any organs of sense may in time grow into an animal with organs of sense"; but "by no effort of the understanding, by no stretch of imagination," he declares, "can I explain to myself how language could have grown out of anything which animals possess, even if we granted them millions of years for that purpose." In other words, he can imagine how a sore spot in the skin could grow into a complex and delicate organ like the ear, or a sensitive black spot could develop into the marvellous mechanism of the eye, but by no mental effort can he conceive how an imperfectly developed convolution in the brain of an ape could become a perfectly developed convolution in the brain of a man. Surely this is one of the strangest freaks of the imagination on record. Yet he admits the correctness of Dr. Broca's conclusions on this subject. "So much," he says, "seems to be established: if a certain portion of the brain on the left side of the anterior lobe happens to be affected by disease, the patient becomes unable to use rational language; while, unless some other mental disease is added to aphasia, he retains the faculty of emotional language, and of communicating with others by means of signs and gestures." This statement is not exact. Aphasia is not the loss of rational language, but of articulate speech, which is something quite different. The aphasiac can exercise his reasoning powers and can entertain and express by pantomime rational ideas, but he is unable to utter or embody them in either oral or written words, although he may understand them when addressed to his ear or eye.

Sometimes there is not an entire cessation, but a curious and comical perversion of speech in the patients, who use words having no connection with the ideas they wish to convey, and are often, though not always, unconscious of any discrepancy or impropriety in their language. Thus Trousseau narrates the case of a lady who, on receiving a call, met her visitor with a kindly smile, and, pointing to a chair, exclaimed, "Pig, brute, stupid fool!" "Madame begs you to be seated," said a friend who was present, and thus interpreted the courtesy really intended by the rude greeting. The lady's conduct was otherwise sensible, and her process of thought logical and rational, although her utterances were wholly irrelevant, and usually most coarse when meant to be most charming.

Another striking case, recorded by Trousseau and cited by Bateman, is that of Prof. Rostan, who, while occupied in reading one of Lamartine's literary conversations, began to be aware that he only partially comprehended the sense of the text. He stopped for a moment, then resumed his reading, and again experienced the same difficulty. He became alarmed, and wished to call for assistance, when, to his surprise, he found himself unable to speak a word. It now occurred to him that he might have had a stroke of apoplexy, but he could move all his limbs and could discover no evidences of paralysis. He rang the bell, but when the servant appeared he could not tell what he wanted. He could move his tongue in all directions, and seemed to have full control of his vocal organs, but could not express a thought by speech. He made a sign that he wished to write, but when pen and ink and paper were brought, although he had the perfect use of his hand, he could not express a thought by writing. After the lapse of two or three hours a physician came, and Rostan, turning up his sleeve and pointing to his arm, thereby manifested the desire to be bled. No sooner was this done, and the local pressure on the brain relieved, than he was able to utter a few words, and after twelve hours was completely restored and could speak as well as ever.

An orang-outang that had once been bled on account of illness, not feeling well some time afterward, went from one person to another, and, pointing to the vein in his arm, signified plainly enough that he wished the operation to be repeated. In this instance, the orang, not being endowed with articulate speech owing to the rudimentary condition of a convolution of the brain, expressed his ideas just as the Frenchman did, who had been temporarily deprived of the faculty of articulate speech owing to the suspension of function in the same convolution of the brain. The process of reasoning was identical in both cases. The idea of recovery from sickness was associated with the act of venesection as the result of experience. In short, the man reverted for the time being to the condition of the monkey. How then should it be deemed a thing impossible for him to have risen out of such a condition?

It is also interesting to note that an injury to the brain of the lower animals sometimes produces phenomena analogous to those of aphasia in man; causing birds, for example, to sing their notes wrong, reversing the intonation and accent, like the quail mentioned by Dr. Abbott, which, owing to such an accident, persistently whistled "white-bob" instead of "bob-white."

It would be superfluous to multiply instances of the

capability of understanding articulate speech manifested by monkeys, horses, dogs, cats, elephants, birds, and other animals which acquire this power, as children do, through the ear and by the exercise of attention. They also show a nice discrimination in distinguishing between words similar in sound. A parrot or a raven masters a new sentence by repeating it, and working at it, just as a schoolboy solves a hard problem. These birds associate sounds with objects, and thus invent names for them. Every dog is a " bow-wow," and every cat a "miau-miau." The denotative term has an onomatopoetic origin, and by the process of generalization is applied to all animals of the species; it is not necessary that the parrot should have heard each individual dog bark or cat mew before giving it its appropriate name. A raven belonging to Gotthard Heidegger, a clergyman and rector of the gymnasium in Zurich, was constantly picking up words dropped in general conversation, and using them afterward in the most surprising manner.

Even animals whose laryngeal apparatus is not structurally adapted to the production of articulate sounds may be taught to utter them. Leibnitz mentions a dog which had learned to pronounce thirty words distinctly. In the Dumfries Journal of January, 1829, an account is given of a dog which called out "William" so as to be clearly understood; and Mr. Romanes cites the case of an English terrier which had been taught to say, "How are you, grandmam?" The careful and systematic experiments now being made in this direction by Prof. A. Graham Bell and other scientists are exceedingly interesting, and may lead to important results.

In view of these facts, it is evident that the barrier between human and animal intelligence, once deemed impassable, is becoming more and more imperceptible, and with the rapid progress of zoöpsychological research will soon disappear altogether. "When we remember," says Prof. Sayce, "the inarticulate clicks which still form part of the Bushman's language, it would seem as if no line of division could be drawn between man and beast, even when language is made the test." Apes make use of similar clicks for a like purpose, and these sounds are doubtless survivals of speech before it became distinctively articulate.

Max Müller expresses great contempt for what he calls "nursery philology," which, he thinks, can throw no light on the origin of human speech. "The two problems, how a child learns to speak English, and how language was elaborated for the first time, are as remote from each other as the two poles." This remark betrays an utter misconception of the objects to be attained by observing the earliest stages in the mental development of infants. It is not to see how a child learns English or German or any other known language, but how it attempts to construct a language of its own for the expression of its thoughts, that interests the psychologist, and may aid him in solving the problem of the origin of speech. That it "can be solved by a careful analysis of language, such as it exists in the immense variety of spoken languages all over the globe," is highly improbable, and, indeed, from the very nature of the questions involved, quite impossible. All efforts that have been made in this direction from the dawn of philosophical speculation to the present time have failed, and are forever predoomed to failure.

The philologist can no more explain the origin of roots by the study of languages than the physiologist can discover the origin of protoplasm by dissecting vital organisms, or the psychologist can determine the origin of the will by appealing to consciousness. Given roots, and there is no mystery about the growth and structure of language; given protoplasm, and the evolution of organic life in harmony with its environment is perfectly intelligible; and a correct conception of volition is the safest clew to the intricate maze of human conduct and the soundest basis on which to build a system of ethics. It must be remembered, too, that in spoken and written language there are no roots, but only the outgrowths of roots—namely, words arranged in sentences. The dictionary defines a root as "a primitive form of speech, one of the earliest terms employed in language"; but so far as our knowledge extends, no tribe of men, however primitive, ever used such a form of speech, and no such terms are found in any language. Roots as such have then no real and independent existence, and, while it would be hardly correct perhaps to call them fictions of the philologist, they are the products of philological analysis, and exist in human speech only as the protoplasmic element out of which it is evolved.

In a brilliant lecture on Atoms, the late Prof. Clifford describes the movements of these molecules as they swing around and then fly away in different directions, mutually approaching and receding, and behaving to one another "somewhat in the same way as two people do who are dancing Sir Roger de Coverley." An enthusiastic critic, who quotes the whole passage, remarks, "Such scientific exposition as this is as beautiful as poetry"; and he might have added, like poetry, it is purely a creation of the imagination. The particles, whose motions are so minutely and vividly depicted, are wholly hypothetical, and, if a matter-of-fact pupil were to ask the professor where he sees all these beautiful things, he could only reply, with Hamlet, "In my mind's eye, Horatio." They belong to the realm of mythopoetic fancy, as truly as do the Muses of Apollo or "the rosy-bosomed Hours in fair Venus' train."

Equally remote from human experience and impossible at any period in the evolution of the human race is Max Müller's often-quoted description of the processes by which such roots as *pat* and *mar* were "tuned down" into their present "smooth and manageable" shape, and rendered serviceable as signs of general concepts. There is no stage in the growth of human speech in which one can conceive of such a process having taken place, and every attempt to imagine and to describe it leads logically to no end of philological contradictions and psychological absurdities.

In the first part of this chapter we quoted Max Müller's theory of "roots as ultimate facts," and his warning of some mysterious danger that would be incurred by the rejection of this view. He now adopts the hypothesis suggested by the late Prof. Noiré, "that the primitive roots of Aryan speech may owe their origin to the sounds which naturally accompany many acts performed in common by members of a family, a clan, or a village." Here we have the strange spectacle of men so highly civilized as to be living together in families, clans, or villages, and yet by their joint efforts in performing certain tasks that require their united strength, creating the primitive roots of their language. But as all words are derived from roots, and all human languages are evolved out of roots, and every thought that ever crossed the mind of man can be traced back to a few simple concepts embodied in roots, what was the language of the families and communities thus engaged in unconsciously, but not the less really, producing the roots of their language? The only logical inference from the premises is that they were speechless, and although Max Müller distinctly and disdainfully repudiates the *Homo alalus*, and declares that he knows nothing of such a creature, the theory he advocates brings us face to face, not merely with the *Homo alalus*, as a solitary individual, but with socially organized masses of *homines alali*, toiling together and "finding relief in emitting their breath in more or less musical modulation," and thus uttering concomitant sounds which express their common acts and become "the germs of conceptual language" called roots. It was in this way that the concept of rubbing came to "be expressed by *mar*, and that of tearing by *dâr*." The same holds true of the roots *pâ* to protect and *mâ* to form. As the words derived from these roots are necessarily of later origin than the roots themselves we are driven to the conclusion that the Aryans, at the time when they began to live in families, clans, or villages, and before they had shouted in unison at their work and thus created these roots, had no words for death (*mara*), or disease (*mâri*), or rent (*dâra*), or even for father (*pitṛi*) and mother (*mâtṛi*), since the parenthetic words are admitted to be the oldest terms in the Aryan family of languages used to express these concepts or to denote these relations, and could not therefore have been preceded by any others, a conclusion which is a complete *reductio ad absurdum* of the whole symphonic or "synergistic" theory as applied to the origin of human speech.

Again, we are told in a passage already cited that the point in the process of evolution where the animal ceases and the man begins is coincident with the formation of the roots of language, and that these roots constitute an impassable barrier between man and beast. If this statement be correct, we are at a loss to know how to classify zoölogically our primitive ancestors, who had not yet had occasion to perform acts in common, and whom the etymogenetic *clamor concomitans* had therefore not yet provided with the stock of radicals essential to the development of speech and of general ideas. So far as their claims to humanity are concerned, they are certainly on the wrong side of the barrier, and can lift themselves over it only by united and persistent exertions of the lungs in crying yo-he-ho!

That there was a time "when the first sound of language burst forth from the breast of the first man, as yet dumb" is admitted by Max Müller, who quotes with approval a sentence to this effect from Steinthal. Here we are again thrown into the disreputable society of " the *Homo alalus*, the speechless progenitor of *Homo sapiens*," notwithstanding the cynical reproach made to Prof. Romanes for seeming to be "so intimately acquainted" with that questionable individual. What sort of creature could "this first man, as yet dumb," this unspeakable Caliban, have been? According to the definition, he did not possess the one essential characteristic of man, and must therefore have been a brute "honour'd with a human shape"; and although we are assured that "whatever animals may do or not do, no animal has ever spoken," speech did burst forth from the breast of this primeval animal, and the anthropoid prodigy, rational but not yet orational, suddenly spoke out and became the ancestor of a flippant and garrulous race.

The origin of language, like the origin of life, is so obscure that some thinkers, in despair of discovering it, have been content to let it remain a mystery by declaring it to be a special gift of God, with which man was endowed at his creation or which was taught him immediately afterward by his Creator. In either case, the gift must have been transmissible from generation to generation in order to become the heirloom of the race. No one, however, will maintain that a child inherits its language from its parents; what it inherits is the faculty, which its earliest progenitors must have also possessed, of producing a language, and this power is not the less creative, because with the child of to-day the process is facilitated and the result determined by its social and domestic environment. But the existence of such a faculty and its continued exercise can be fully accounted for by the doctrine of evolution without necessitating the intervention of a *deus ex machina*. Indeed, an act of special creation would explain or rather account for the possession of this faculty by the person on whom it was miraculously conferred, but not by his descendants. As Steinthal observes, "What a man has been exceptionally endowed with by God no other man can learn from him." Only that is learnable which comes through the natural and progressive development of the power of learning inherent in the race, and which each individual is capable of learning for himself, though less easily than through intercourse with others. The theory of the divine origin of language may therefore be set aside as unscientific, since it evades the question instead of answering it, and complicates the problem by substituting two mysteries for one, and as inadequate, since it fails to account for the phenomena in their historical continuity.

Apart from the untenable assumption of supernaturalism, all principles involved in the origin of language may be reduced to three : the onomatopoetic, the interjectional and the synergistic principle, or according to Max Müller's vernacular and expressive nomenclature, the bow-wow, the pooh-pooh, and the yo-he-ho theories. " Those who appeal to words like thunder as derived from the rumbling sound in the clouds without any conceptual root standing between our conceptual word thunder and these unconceptual noises, hold the bow-wow theory. Those who hold that fiend is derived from fie, without any conceptual root standing between the unconceptual fie and the conceptual word fiend, hold the pooh-pooh theory. Those who would derive to heave and to hoist from sounds like yo-he-ho, would hold what may be called the yo-he-ho theory." In this connection, Max Müller states that " the yo-he-ho theory is the very opposite of what Noiré called the synergistic theory "; although he does not make the distinction clear and was himself the first who substituted this slang term for the dignified Greek designation, when he endeavoured to show how the roots *mar*, *dâr*, and *tan* might have been produced by men engaged in common acts of grinding, tearing, and stretching, and finding relief in emitting their breath in musical modulation, as sailors are wont to do in pulling ropes. The same synergistic activity of primitive men in gulping their food and the noise of simultaneous deglutition would give rise to the root *gar*, to swallow; whereas the reverse process of disgorging the contents of an overloaded stomach and the unmistakable sound attending this operation, especially when performed by several persons synchronously, as was the custom of Roman gormands at their banquets, would be expressed by *vam*, a root that is found in every Aryan tongue, thus proving at what a very early period this vigorous race began, as the Germans say "to call upon the name of St. Ulrich." It will never do for the godfather of Yo-he-ho, who stood sponsor for the infant and held it so tenderly in his arms at the font, to repudiate this youngest born of philological bantlings, simply because, as it grows older, it bears so strong a resemblance to its big and burly brother Bow-wow.

As regards the second of the above-mentioned theories, we can hardly believe that even the most inveterate pooh-poohist would derive fiend from fie either directly or indirectly. Irascible Germans, under strong excitement, are apt to link the words together in the scornful phrase, *Pfui! Teufel!* But no one who has given any thought to the subject would connect them etymologically.

In the sentence which we have quoted there seems to be also a queer confusion of ideas concerning the conceptual and the unconceptual. Thunder, regarded as the report which follows the discharge of atmospherical electricity, may be properly spoken of as an unconceptual noise; yet it is something more than this to the primitive man or the savage, who hears in it the voice of an angry deity. Again, if a person wishes to inform another that he has heard thunder by pointing to the sky and imitating the sound, the rumble thus produced is no longer unconceptual, but becomes the sign of a distinct and intelligible idea. So, too, of interjections, such as fie and fudge, they express the concept of mingled incredulity and contempt as clearly as any collocation of words in a sentence could do. This theory is carried so far by its author as to lead him to the assertion that mispronunciation annihilates a word, the change of a vowel or consonant, or "only an accent" sufficing to deprive it of its articulate character and to resolve it into empty noise, or "what Heraclitus would call a mere psophos." This position is justly characterized by Prof. Whitney as "not only wrong, but ludicrously wrong." What becomes of the hundreds of words in the English language which many persons habitually mispronounce, and even lexicographers accent differently? Does this deviation in orthoëpy destroy the conceptual quality of the word and reduce it to a mere noise? As regards the term psophos (ψόφος), here somewhat pedantically introduced, it was used by Greek writers in distinction from phônê (φωνή), vocal sound or tone, and Aristotle calls *phônê* the *psophos* of animate creatures (ἡ φωνή ψοφος τίς ἐστιν ἐμψύχου), but it did not necessarily mean mere noise. It was also employed to denote the cries of people in the street or on the market place, and other significant sounds, including those produced by insects as well as the conventional rap on the door (ψοφεῖν τὴν θύραν) by which a visitor announced his presence and virtually asked whether he might enter. On the other hand, phônê was applied to the utterances of beasts as well as of men, and even to the noises of inanimate things, as when Sophocles speaks of the voice of the loom (κερκίδος φωνή). Indeed, if any inference could be drawn from an exhaustive and critical comparison of the passages in which these words are used, not only by Heraclitus, but in a far more frequent and philosophical manner by Aristotle and his contemporaries, it would certainly not favour the notion that a word is robbed of its significance by mispronunciation, or converted into mere noise by a false accent. That epochal *clamor concomitans* in which, according to the synergistic and, as Whitney calls it, "utterly fantastic theory," language had its origin, would have been to the Greeks a mere psophos.

Wholly untenable, too, is the assertion that no traces of conceptual thought are discoverable in the lower animals. The very lowest forms of organic life communicate with each other by means of sounds, although in many cases the manner of their production has not yet been definitely determined. Some insects possess vocal organs in the proper sense of the term and are thus enabled to give voice to their emotions, while others express their desires in a more mechanical way by the buzzing vibration of their wings or the stridulous friction of their legs. The death's-head moth has a sort of bagpipe arrangement, consisting of an internal sack and a proboscis through which the air is forced, producing a shrill and doleful treble like that of the so-called chanter of the Scotch instrument. Gnats have a twofold mode of expression—a voice accompanied by deeper tones made by the movement of the wings. This combination of sounds constitutes their language and, in a limited degree, appeals to them in the same manner as speech does to man, so that they are attracted by an imitation of it with the human voice or on a violin. Prof. Landois gives a comical example of this in his work on Thierstimmen: "One day," he says, "I found my servant boy in the garden engaged in his favourite occupation of doing nothing. By chance a swarm of musquitoes was hovering in the air near me. I called the fellow and reproved him for his laziness, and raising my voice to the musquito pitch of high E said: 'If you don't go and black the boots properly, I'll have you stung to death by musquitoes.' The words were hardly uttered when the whole swarm came down upon him, causing him to flee in terror from the wizard, who had even the musquitoes at his command." In this purely musical language the tone varies slightly perhaps with the individual, but very perceptibly with the sex of the insect, the male having a somewhat higher note than the female. With every advance in the ascending scale of animal life the power of expression increases, gradually ceasing to be a monotonous hum or dull drone, and becoming considerably modulated and growing more and more articulate until it reaches its highest development in human speech. Great as may be the disparity between the squeak of a mouse, the chatter of a parrot, the roar of a gorilla, the gibberish of a Bushman, and the eloquence of a Demosthenes, there is really no break in this long process of evolution corresponding to the growth of the intellectual faculties in the several species. Every creature has a language of its own composed of significant sounds intelligible to its kind; and there is no point in the development of vocal utterance at which it can be said hitherto these sounds have been empty and unconceptual noises, henceforth they express the wants, emotions, and ideas of those who use them.

Mr. James Weir, Jr., to whose special study of the senses of the lower animals reference has been made in a former chapter, says that, although ants are generally supposed to communicate with one another through their antennæ, they may do so by means of sounds so low as to be inaudible to the human ear, but adds that he has never been able to find in them either vocal organs or instruments of stridulation, such as the genus *Gryllus* possesses. The recent microscopical observations of M. Charles Janet, however, have led him to the conclusion that ants are provided with vocal organs, and he claims to have succeeded in ascertaining by the use of the microphone that they give utterance to a considerable variety of tones, which he assumes to be expressive of different emotions.

From a philological point of view Max Müller has as slight consideration for beasts as for babies, taking every occasion to disparage the teachings of zoöpsychology, and refusing, as he says, to argue with any philosopher "either in the nursery or in the menagerie." Curiously enough, it now seems as though the possibilities of the menagerie in this direction had been strangely overlooked, and the ultimate elements of human speech, "phonetic cells," might yet be discovered in the monkey's cage. About a dozen years have elapsed since Mr. R. L. Garner began his study of the language of quadrumans in the Zoölogical Garden at Cincinnati.* Confined in a large cage with a number of smaller monkeys was a mandril endowed with all the ugly characteristics of his kind. This beast, which in its wild state is spoken of by the natives as the "forest devil," and which science has

---

* Although Mr. Garner's researches have led to no definite results, and he has been rather harshly characterized as "a sensational charlatan," I prefer to let my remarks on his attempt to solve the problem of simian speech remain as they were originally written, since they could not be eliminated without impairing the general discussion of the subject.

named the "dog-headed horror" (*Cynocephalus mormon*), and whose habits of life justify both designations, is the most hideous in appearance, the most cruel in disposition, the lewdest and wickedest of baboons. Perhaps in no other creature is "pure cussedness," or the love of evil for evil's sake, so highly developed. It is the only animal, except man, that undermines its health and dies a premature death through excessive and unnatural indulgence of its lusts. Jardine reports the case of a mandril that, in becoming domesticated, took readily to the vices of civilization, brandy-drinking, and smoking, although it greatly preferred alcohol to tobacco. Broekmann, however, succeeded not only in taming one, but also in overcoming its vicious propensities by a proper course of training. He taught it a variety of tricks, which it had to perform every day in an orderly manner and gradually took pride in performing well, thus proving that if "idleness is the parent of vice," regular industry is the source of sobriety. The distinguished naturalist Reichenbach, who watched Broekmann's experiment with lively interest, was struck by the wonderful effect which the mere fact of having something definite to do produced in transforming the wildest and most wanton of baboons into a decent and quite companionable creature. As the result of this systematic discipline, "its lower and purely animal propensities and carnal appetites, which tended to undermine its own existence, began to calm, and ceased to be easily excited as its higher faculties were awakened and called into exercise, and as it was drawn upward through instruction and through love of the feats whose performance had kindled in it the first spark of mental activity and now kept its powers in a constant state of tension." Here we have an application of the same means that are used to civilize a savage or to train up a child in the way it should go; and Broekmann's success in curbing the cruel mandril of its will, elevating it socially into an agreeable and affectionate friend, and educating it histrionically to be the "star" of a monkey theatre, is not only a triumph of pedagogical patience and skill, but also a striking proof of the animal's latent capabilities.

The mandril at Cincinnati was a savage brute which killed several of its less powerful associates and made the lives of the rest hardly worth living. They were never free from alarm and always on the alert to escape the assaults of their common foe, whose movements they anxiously watched and quickly reported to each other. Mr. Garner, who observed them day after day, soon became convinced that their cries and chatterings were not mere unconceptual noises or ejaculations inspired by individual fear, but vocal expressions of ideas conveying definite information. He endeavoured to start a conversation by imitating these sounds, but the monkeys, although their attention was somewhat attracted by his utterances, evidently failed to comprehend his broken Simian, and in a short time ceased to pay heed to a creature who could not talk better than that. It then occurred to him to use the phonograph, which would not only reproduce these sounds with precision, but would also repeat them at pleasure and thus enable the human voice by persistent practice to articulate them distinctly. This plan was successful, and the monkeys showed by their actions that they clearly understood what the phonograph said. It is not necessary here to enter into the details of his subsequent experiments, the results of which he has embodied in a volume; and although one may not accept some of his inferences, his researches all tend to confirm the assumption that monkeys have a language of their own, and that man can learn it and converse with them in their own tongue. Mr. Garner is now in the wilds of Africa, amply equipped with the instruments requisite for his personal safety and for the pursuit of his investigations in the habitat of the gorilla and the chimpanzee; and the fruits of the studies which he is carrying on under so favourable circumstances will be awaited with interest. *Nomen et omen:* may he return with a well-filled garner!

Brehm remarked nearly twenty years ago: "The language of apes may be called quite rich; at least every ape has at its command a great variety of tones for the expression of different emotions. Man also learns the significance of these sounds, which are difficult to describe and still more difficult to imitate." Indeed, without the aid of the phonograph it would have been impossible to determine their exact nature and to reduce them to a phonological system. This is an interesting illustration of the far-reaching and beneficent influence of great inventions. The phonograph may yet render as valuable service to philology by extending the field of linguistic research as the microscope has rendered to medicine, and especially to bacteriology.

It has been repeatedly asserted and generally accepted that "no animal has the power of forming roots," and that there is "not one root in the language of animals." This statement is sheer assumption, and for aught we know the very reverse of the proposition may be true and the language of animals consist chiefly, if not wholly, of roots. The origin of these constituent elements of language is a mystery. No philologist can tell how they arose or why they should convey one meaning rather than another, except so far as their genesis may be explained in a few cases by onomatopoetic suggestion. The most natural supposition is that they are the heirlooms of the anthropoid race, which may have been transmitted to man from the semi-human primates of the Miocene age, already highly specialized enough to be capable of chipping flints. The beauty and force of human speech and its superiority to the utterances of brutes are due, not to the fact that it has its origin in roots, the very existence of which is overlooked by the great majority of men and detected only by philological analysis, but to its marvelous growth out of roots, to its grammatical and syntactical structure, its elaborate and complicated system of cases and tenses, the etymological relations of its parts, and the various means employed to express the nicest shades and most subtile suggestions of thought and feeling. It is this wonderful and never-ceasing evolution that makes human language what it is and distinguishes it from the extremely scanty and comparatively stationary language of all animals from the death-watch to the *Dryopithecus*.

Reasoning from what we know of the language of apes, that of "the missing link" must have consisted mostly of monosyllabic sounds expressive of simple concepts, which, in proportion as his posterity reached a higher degree of intellectual development and became partially humanized, were gradually modified in meaning by prefixes and suffixes and organically correlated by inflection, until the original monosyllabic utterance, often perhaps little more than a short and sharp outcry, ceased to be used except in these derivative and differentiated forms. The root was thus merged and wholly lost sight of in the word, as is now the case with all human tongues, where its existence is unsuspected, until the philologist grubs it up. No one maintains that human language was evolved in this manner out of that of the gorilla or chimpanzee or any other existing tribe of *Simia*. As the different races of men are descended from different anthropoid species now extinct, so the diversity of tongues arose from the same cause. These anthropoid species or types died out because they were too nearly akin to man to compete with him in the struggle for existence; they could survive only by sharing in the advance of the race, and by sharing in it they ceased to be what they were; in either case they were doomed to disappear.

In this connection we may refer again to the assertion of the identity of language and thought, or the statement that reasoning can not be carried on without words. If this principle be correct, it is rather queer that the most thoughtful persons should be, as a rule, the least wordy, or as Shakespeare says of Nature, "deep clerks she dumbs," as though it were one of her universal laws. Emotionally there are, as Wordsworth tells us, " thoughts too deep for words"; intellectually there are thinking processes, such as the abstract consideration of quantity and magnitude and their relations in pure mathematics, for the expression and realization of which words are far too clumsy and inexact, and recourse must be had to figures, algebraic characters, and arithmetical and geometrical formulas. On the other hand, there is a low stage of thought, which a few gestures or exclamations are fully adequate to represent. Indeed, in Old Chinese, the archaic Kuan-hua, we have an extended literature recorded in what is essentially a language of signs, intelligible to the eye but not to the ear, and having no grammatical connection or any relation to each other except propinquity. Sinograms, ideograms, and all hieroglyphics and picture-writing are relics of this early period in the evolution of alphabetical language.

The beast, as well as man, is not confined to the use of what might be called its native tongue, but is capable of learning foreign languages. Not only do parrots, ravens, and other birds acquire considerable facility in human speech, but animals, in which the social instinct is strongly developed, adopt the means of communicating thoughts employed by other animals with which they habitually associate. A striking example of this adaptation is given by Dr. Paul Carus: "If ants of a special kind rob the larvæ of another kind and educate them as their slaves, the slaves will in case of war or danger stand by their masters even against their own folks. They evidently speak the language of the hill in which they have been raised," just as children carried off in their infancy speak the language of the tribe in which they have been reared, and indeed as all persons speak the language of the community in which they have grown up.

The question whether the roots of language ever existed by themselves or whether any language could consist solely of roots may be "a foolish question" to the philologist, who does not dare to go beyond them; it is certainly a question which his methods will never solve. Less than a century ago there were eminent scholars who regarded the study of Sanskrit as a vain pursuit, and some denounced the language itself as a fabrication of cunning Brahmans. Even as late as 1820 the distinguished Orientalist Silvestre de Sacy wrote to Bopp, urging him to abandon this study as having no immediate relation to classical or theological studies; to-day every one knows that it has revolutionized these studies. The philologist no longer devotes himself exclusively to classical languages or even to those of civilized nations, but extends his researches to the dialects of barbarous tribes. Nevertheless there are still those who refuse to enlarge the borders of their science so as to include any of the lower animals, however intelligent, and obstinately reject all contributions from this quarter. Yet if we ever discover the origin of roots, it will probably be by searching for them in this direction. Least of all is it befitting an earnest and broad-minded scholar to treat such investigations with facetious flippancy, and to repudiate the conclusions of the zoöpsychologist without having any profounder knowledge of the mental powers of animals than can be obtained by walking through a museum and contemplating the dry hides to which the taxidermist has given the semblance of life.

The enthusiasm with which Mr. Garner has devoted himself to the study of simian speech, and the general interest excited by his discoveries, naturally suggest a comparison of his investigations with those of his predecessors in this department of linguistic research. Perhaps the most serious and scientific attempt of this kind was made nearly a century ago by Gottfried Immanuel Wenzel, who published at Vienna, in 1808, a volume of 216 pages entitled Neue auf Vernunft und Erfahrung gegründete Entdeckungen über die Sprache der Thiere (New Discoveries concerning the Language of Animals, based on Reason and Experience), in which he maintained that the lower animals are capable of expressing their thoughts and emotions by means of articulate sounds, and that these utterances are not only intelligible to their kind, but may also be understood by man, indicated by alphabetical signs, and thus reduced to writing. He made a list of the sounds uttered by thirty different birds and beasts, and prepared a dictionary of more than twenty pages, to which he added a number of translations from animal into human speech. These so-called translations are very free, and give merely a paraphrastic statement of what he supposes to be the significance of certain canine and feline tones, the versions being confined to his interpretations of the colloquies of cats and dogs. As an illustration of his proficiency in this language and the practical value of such knowledge, he relates an incident, which sounds as though it might belong to the ancient and fabulous literature known to the Germans as *Jägerlatein*, or hunters' Latin. He once went to visit a friend, who was a great huntsman, but on learning that he had gone out with his gun waited for him to return; meanwhile he took a book and sat down under a tree near a pen in which some foxes were confined. Suddenly he heard them utter certain sounds which according to his vocabulary were expressive of surprise and joy, and after listening for a time came to the conclusion that the foxes had discovered some means of escape and were exulting over the prospect of regaining their freedom. When the hunter returned, Wenzel informed him of what he had heard and advised him to look into the matter, but was only laughed at for his credulity and assured that the pen was perfectly secure. They went into the house, where they were taking some refreshments and talking about other affairs, when a servant rushed in greatly excited and announced that the foxes had escaped.

Wenzel admits that the language of animals is extremely simple and limited, and consequently monotonously repetitious; the same combination of sounds uttered with a stronger or weaker intonation serves to denote a variety of mental states and must be largely supplemented by lively pantomime. In conclusion, he has eighteen pages of what he calls an "animal pathognomonic-mimetic alphabet," showing the value and function of each part of the physical organism, from the teeth to the tail, as a vehicle of expression. Dogs and cats fairly bristle with strong emotions, and birds show their ruffled feelings in their feathers and wax eloquent with their wings. Wenzel is convinced that every species of animal has its own dialect, which is to be regarded as a modification of the common or generic language of the race to which it belongs. Thus he seems to think that the zebra would understand the ass more readily than the horse, because the first two are more closely affiliated, although all three are endowed with equine speech. The same principle applies to the different varieties of the domestic hog in relation to other suilline quadrupeds.

As an example of the extent to which animals may acquire a knowledge of human speech he prints a communication from a clergyman who had taught his dog to fetch books from his library in an adjoining room. "Fido," he would say, "on the table near the window are a quarto, an octavo, and a duodecimo; go and get the quarto." Fido never failed to bring the volume designated. He had trained the dog to perform this service by showing him a book and saying very distinctly and repeatedly quarto, octavo, or duodecimo, and then laying it down in the library and making him fetch it. In the same manner the dog was taught to bring many other objects, the names of which he seldom confounded or misunderstood. The clever animal could also be sent on errands. "Fido," the clergyman would say, "go to Mr. B. and tell him that I shall call upon him to-day." Thereupon Fido ran to Mr. B.'s house and on finding him gave three short barks, which were perfectly intelligible to the person thus addressed. If any one called when the clergyman was out, Fido barked once; and he did the same if his master did not wish to be disturbed and bade him tell the caller that he was not at home. He announced a visitor by scratching on the door and barking twice. A Bavarian family at Munich has a dog that deems it highly improper for gentlemen to wear their hats in the house, but is sufficiently gallant not to find fault with ladies for doing so. An American, who wished to test the animal's discriminating sense of the fitness of things in this respect, entered the room and sat down with his hat on. The dog looked at him disapprovingly for a moment and then began to bark, with eyes intently fixed upon the hat. As the unmannerly visitor continued the conversation without paying any attention to these admonitions, the dog sprang up and, seizing the hat by the brim, pulled it off and quietly laid it on a chair.

Wenzel also tells the story of a dog whom his master used to send to the market for meat, and who would stand before the kind of meat he was instructed to get, beef, mutton, or veal, and bark once, twice, or thrice, according to the number of pounds desired. The butcher filled the order, and the dog trotted home with his purchase and the cheerful consciousness of having done his duty. A still more remarkable case of this kind occurred recently in a German town, where the dog went regularly to the market and always to the same stall. One day the animal took offence at something and immediately transferred his custom to another stall. A few weeks later the owner of the dog met the first butcher and remarked that the meat, although a little cheaper, was not quite so good as formerly. "There must be some mistake," was the reply, "for I have not sold you any meat for a long time." This statement led to an investigation and final solution of the mystery, which created much amusement and excited no little astonishment that canine minds could "such high resentment show." Wenzel's little book is full of interesting anecdotes illustrating his subject, and has a frontispiece representing a landscape, resembling the traditional pictures of the garden of Eden found in old Bibles, with an ape, a dog, a horse, and a bull in the foreground, and the legend underneath: "They do not lie; their speech is truth."

The French physicist, R. Radeau, in a work on acoustics, published in 1869, treats incidentally of the language of animals, which he thinks one could, by careful observation, learn to understand and even to speak with fluency. Mersenne, in his Harmonie Universelle, asserts that men speak from a volitional impulse and utter vocal sounds in the exercise of a power of the mind which they are free not to exercise unless they choose to do so, whereas the lower animals use their voices under the influence of natural necessity, howling, shrieking, singing, etc., because under the circumstances they can not do otherwise, being subject to forces which they are absolutely unable to resist. The vexed question of the freedom or necessity of the will in human action, which metaphysics has vainly endeavoured to solve, has been reopened by natural science and evolutionary biology and is now discussed on a broader basis and with the prospect of positive results. Whatever may be the final issue of these investigations, it is certain that the old Cartesian distinction between man and brute in this respect can no longer be maintained. Radeau is right in rejecting Mersenne's theory as involving a too subtle psychological distinction and in declaring that his doctrine of natural necessity might be applied with equal force to many an inveterate gabbler who can not hold his tongue.

In this connection he relates the following anecdote on the authority of Jules Richard: In 1857 this gentleman had occasion to visit a sick friend in a hospital, where he made the acquaintance of an old official of the institution from the south of France, who was exceedingly fond of animals, his love of them being equalled only by his hatred of priests; he claimed also to be perfectly familiar with the languages of cats and dogs, and to speak the language of apes even better than the apes themselves. Jules Richard received this statement with an incredulous smile, whereupon the old man, whose pride was evidently touched by such scepticism, invited him to come the next morning to the zoölogical garden. "I met him at the appointed time and place," says Mr. Richard, "and we went together to the monkeys' cage, where he leaned on the outer railing and began to utter a succession of guttural sounds, which alphabetical signs are scarcely adequate to represent— 'Kirruu, kirrikiu, kuruki, kirikiu'—repeated with slight variations and differences of accentuation. In a few minutes the whole company of monkeys, a dozen in number, assembled and sat in rows before him with their hands crossed in their laps or resting on their knees, laughing, gesticulating, and answering." The conversation continued for a full quarter of an hour, to the intense delight of the monkeys, who took a lively part in it. As their interlocutor was about to go away, they all became intensely excited, climbing up on the balustrade and uttering cries of lamentation; when he finally departed and disappeared more and more from their view, they ran up to the top of the cage and clinging to the frieze made motions as if they were bidding him good-bye. It seemed, adds Mr. Richard, as though they wished to say, "We are sorry to part and hope to meet again, and if you can't come, do drop us a line!"

No one who has ever observed the actions and listened to the utterances of a clever parrot will accept Mersenne's assertion that the exercise of the vocal organs of animals is not free, but subject to natural and irresistible necessity, or that speech is in a greater degree the product of inevitable causation in the mouth of the cockatoo than in that of the cockney. Humboldt states that after the Aturians on the Orinoco had become extinct, the only creature that could speak their language was a very aged parrot, condemned by adverse fortune to spend the remnant of its days in comparative solitude as the sad survivor of a once powerful tribe. From a philological point of view, the venerable bird was as interesting a character as the old Cornish woman with whose decease, some years ago, the dialect of her people ceased to be a spoken tongue. It is also a historical fact that when, in 1509, the Spanish freebooters Nicuesa and Ojeda wished to surprise the village of Yurbaco, on the Isthmus of Darien, in order to capture a cargo of slaves, the vigilant parrots in the tops of the trees announced the approach of the enemy, and thus enabled the inhabitants to escape.

Perhaps the most cultivated and certainly the most celebrated parrot of which we have any record belonged from 1830 to 1840 to a canon of the cathedral of Salzburg, named Hanikl, who gave the bird regular instruction twice a day, from nine to ten in the morning and from ten to eleven in the evening. The parrot made rapid progress in the development of its mental faculties, and soon showed what a remarkable degree of intelligence it is possible for such a creature to attain under systematic tuition. The sayings and doings of this parrot which lived fourteen years after Hanikl's death and died in 1854, have been reported by a number of careful and competent observers and are unquestionably authentic. One day, as some one entered the room, it cried out in a harsh tone, "Where do *you* come from?" On seeing that the person was an ecclesiastical dignitary, it added, apologetically: "Oh, I beg pardon of your Grace; I thought it was a bird." It took part in general conversation, and was sometimes so loquacious that it had to be told to stop; it was also fond of talking to itself, and imagining all sorts of exciting scenes: "Beat me, will you? Beat me, will you? Oh, you rascal! Yes, yes, that's the way of the world." It whistled tunes and sang various popular songs, and even learned an entire *aria* from Flotow's opera of Martha.

A parrot of the same species (*Psittacus erithacus*), ash-gray, with scarlet-red tail, is now in the possession of M. Nicaise, a member of the Anthropological Society of Paris. This bird is nearly fifty years of age, and endowed with wonderful versatility of intellect. It imitates to perfection all the calls and cries of the street, and when in 1870 it was sent away from the beleaguered city into the country, it came back with its repertory immensely enlarged, having learned to reproduce the whistle of the quail, the hoot of the owl, the merry scream of the magpie, the crow of the cock, the cluck of the hen, and the tones of a great variety of wild birds and domestic fowls and quadrupeds. One of its histrionic masterpieces is the phonetic representation of the killing of a pig which it witnessed nearly a quarter of a century ago, but of which it has not forgotten a single characteristic grunt or squeal. Nothing is omitted, from the deep gutturals, alternating with piercing shrieks, as the porker is dragged to the place of slaughter, to the last faint groan of the dying animal. Indeed, the reproduction of the scene is so intolerably realistic, that the persons present are fain to stop their ears and to bid the bird keep silence. It listens attentively to any conversation that is going on, and expresses its approval or astonishment by exclaiming "Oh!" or "Ah!" and always at the appropriate time or place. If any one tells a funny story or gets off a joke, it laughs with the rest of the company, although this outburst of merriment is doubtless due, not so much to a humorous appreciation of what is said, as to the contagion of the general hilarity.

When it wants something, it calls its mistress by her Christian name, Marie, and, if she does not come at once, calls her again with a sharp tone of impatience. Once, when a firebrand fell on the hearth and filled the room with smoke, it cried, "Marie! Marie!" in a voice indicating extreme anxiety and alarm. This parrot is a provident creature, and when taking its dinner always lays aside a piece of bread and jam for its supper, thus showing that it has the power of looking before and after, which Shakespeare deems a peculiarly human attribute. It not only sings songs correctly, but also improvises musical compositions, which it renders each time with new variations, and performs, as M. Nicaise assures us, " with a taste and style and spirit that might excite the envy of any pupil of the conservatory." The fact that these pieces invariably close on the tonic or keynote proves that all the modulations are referred to the fundamental tone of the chord, and gives evidence of a musical feeling and sense of harmony such as only human beings are usually supposed to possess. These improvisations are whistled, and sound as though they were played by a flute, the performance being uniformly preluded with runs and trills and other vocalizations.

The parrot is an exception to the rule that the period of infancy is longest in the most intelligent creatures. Its babyhood is, in fact, very short, although its average life seems to be somewhat longer than that of a man. It attains the full splendour of its plumage and is pubescent at the early age of two, and often survives all the members of the human family in which it has been reared, outliving even the children much younger than itself. During all this time it retains its mental plasticity and progressiveness, never ceases to learn, and goes on developing its inborn capacities from the beginning to the end of its prolonged existence. It is quite as inquisitive as the monkey, and quite as capable of close and continued observation. Merely through its association with man it is constantly making new acquisitions of knowledge, and there is no telling what might not be accomplished in this direction by systematic instruction carried on through successive generations.

If Mr. Garner's object had been to ascertain how far

animals can acquire the use of human speech and what effect such discipline would have in enlarging their intellectual faculties, he would have done better to choose parrots instead of monkeys for his experiments; but as his purpose is to learn the language of animals, and not to teach them his own, he has done well to select apes as the objects of his study. It must be confessed, however, that the results of his investigations, embodied in his volume recently published, are rather disappointing, and are, in fact, less comprehensive, although doubtless more accurate, than the observations made by Wenzel at the beginning of the present century. He is prone to lay great stress upon matters that are really of no importance whatever, as, for example, when he discovers that "No" accompanied by a shake of the head is the sign of negation, and adds, "The fact that this sign is common to both man and simian I regard as more than a mere coincidence, and I believe that in this sign I have found the psycho-physical basis of expression." It is difficult to perceive how a logical thinker could draw such a sweeping conclusion from so slight premises. If he finds that gorillas and chimpanzees in their native wilds, unaffected by human associations, express dissent by shaking their heads and shouting "No!" it will be a fact well worth recording.

Mr. Garner's superiority to his predecessors in this department of linguistic research consists in the greater excellence of his material rather than of his mental equipment. The possession of the phonograph alone gives him an immense advantage in this respect, by enabling him to record and to repeat the utterances of monkeys with perfect accuracy. Armed with this scientific weapon of phonetic precision and all the instruments and appliances which modern invention has placed at his disposal, he may perhaps completely conquer a province of investigation hitherto but partially explored, and, by making important contributions to zoöglottology and working out a system of alphabetical signs for the language of the anthropoid race, become the Cadmus of the simian world.*

---

* According to the latest reports, Mr. Garner has returned from his expedition without having realized the exalted hopes excited by his elaborate preparations and the somewhat sensational announcement of his programme, published on the eve of his departure in The North American Review. In an address delivered before the Société de Geographie in Paris, on May 4, 1894, the African explorer Dybowski took occasion to refer to Mr. Garner's sojourn in the jungles of the Congo for the purpose of learning the language of the gorillas from their own lips. Dybowski stated that he himself had passed two days at the mission of Fernand Vaz, situated on the shore of the lake bearing the same name. The superior, Father Bichet, informed him that Mr. Garner had spent three months there—not in the depth of the forest, but at the mission itself—evidently preferring the society of the monks to that of the monkeys. Mr. Garner brought with him his famous cage "of steel wire woven into a diamond-shaped lattice," and set it up at a place called Fort Gorillas, on the edge of the forest, just twenty-eight minutes' walk from the mission and within hearing of the church bells. Dybowski expresses a doubt whether "the apes, however strong their instincts of civilization, ever came so near the convent to perform their religious devotions." The negro boy Rozoungé, a youth about thirteen or fourteen years of age, who speaks French very well and accompanied Mr. Garner on his excursions, confirmed the statements of Father Bichet, and added that Mr. Garner had slept three nights in the cage, where he awaited in vain the visits of the chimpanzees and gorillas. The boy thinks they heard them one evening, and that is the extent of Mr. Garner's intercourse with these great anthropoid apes in their wild state. He succeeded, however, in buying a young chimpanzee, which he named Moses, but which soon died. He afterward expended sixteen dollars in the purchase of a young gorilla, which survived only a few days. He then left the mission, where he had paid five francs a day for board and lodging, and set out with Father Buléon, of the Congregation of the Holy Ghost, on a tour to the Eschiras, a tribe of the interior. After two days' travel he was taken with a severe pain in his legs, and had to be borne in a hammock to the Tomlinson factory, where he remained two months. On recovery he embarked for Europe, taking with him his cage and the elements of his dictionary of the simian tongue. Dybowski says that the phonograph which was to catch the sounds uttered by the apes and to record them on a cylinder never arrived, so that Mr. Garner had to carry on his investigations without the aid of this instrument. Dybowski's remarks on this subject were reported in the Paris Figaro, and have been published in other papers—e. g., in The Nation, June 14, 1894. The French journal characterizes Mr. Garner as an amateur in linguistics, who has succeeded in having himself taken seriously, but who proves to be "*un simple fumiste.*" It is hardly possible that these statements should have escaped Mr. Garner's notice, and the fact that he has made no reply to them is a tacit admission of their correctness. This utter failure to accomplish what he set out to do is greatly to be regretted, inasmuch as the line of his researches was in the right direction, although they might have been more successfully pursued in a zoölogical garden than in the wilds of Africa or at a missionary station on Lake Fernand Vaz.

Those who may have condemned Dybowski's strictures as too severe will find them fully justified by Mr. Garner's own relation of his experiences and observations recorded in his recently published volume Gorillas and Chimpanzees. The only original discovery he seems to have made is that of the armadillo, hitherto supposed to be peculiar to South America. We venture to assert that the one he saw will probably prove to be the sole specimen of the *Dasypus sexcinatus* existing in tropical Africa. His work does not contain a single noteworthy contribution to simian speech. Indeed, it is difficult to determine the exact field of his labours by consulting any map of the country, and the fact that letters were delivered to him in his cage would imply that the postman was abroad, and indicate that his camp was not far from the outskirts of civilization.

In a communication to an English paper on his return from

Africa Mr. Garner speaks of conferences with gorillas, and adds: "My preliminary understanding of the sounds uttered by my anthropoid visitors was that their government is strictly patriarchal and that they have some fixed idea of order and justice." This theory of the organization of the simian horde is not new, but it is the first time that positive information on the subject has been received directly from the mouths of the *Simia* themselves. The gorillas must have been in a very confiding mood, or they would not have imparted this knowledge to a perfect stranger. Perhaps their communicativeness was the result of hypnotic suggestion, for in a letter published in an Australian journal, the Sydney Daily News, Mr. Garner states that on one occasion he placed his battery with a phonograph and a revolving mirror in a banyan grove and concealed himself about sixty metres distant. A crowd of chattering monkeys soon gathered round the glittering mirror. Mr. Garner observed them for more than an hour and then emerging from his hiding place cautiously approached. No sooner did they see him than they all disappeared as by magic with the exception of one chimpanzee, which stood perfectly still, staring at the mirror, while a slight tremor ran through its limbs and its ears gave a convulsive twitch. "I could hardly believe my eyes; the monkey was hypnotized." As the chimpanzee kept saying "*achru*." which means sun in the anthropoid tongue, one might suggest that this animal was a sun-worshipper in an ecstasy of devotion at the supposed descent of its god to the earth. There is no limit to hypotheses in such cases. Mr. Garner's mention of the phonograph can be reconciled with the positive statements of Dybowski and the French missionaries only by assuming that this apparatus arrived after he left Fernand Vaz.

CHAPTER IX.

THE ÆSTHETIC SENSE AND RELIGIOUS SENTIMENT IN ANIMALS.

The æsthetic sense as a distinction between man brute. Man's artistic faculty the mark of his pre-eminence according to Wilks, Huxley, Prantl, and Schiller. Herbart recognises no such line of separation. The influence of infancy. Value of flexible organs of prehension. Appreciation of the beautiful shown by birds. Fondness of finery. Decorative taste of the bower bird. Nests of the weaver, oriole, titmouse, and japu. Love of music in birds and insects. Musical training of unmusical birds. Beethoven and the spider. The cicada as a violinist. Musical performances of apes in their native wilds. Exhibition of musical preferences by dogs. Musical concerts by mammals and birds. Hudson's observations in La Plata. Propagation of plants dependent upon a sense of colour in insects. Religious sentiment of animals recognised by De Quatrefages and Darwin. Fetichistic conceptions formed by the higher animals. Striking examples given by Herbert Spencer and Romanes. Sense of the supernatural in horses and dogs. A haunted canary cage. Second sight and ghost-seeing attributed by popular belief to dogs, horses, and storks. Religion as a natural growth has its roots in animal intelligence.

DR. WILKS reduces the chief difference between man and brute to the " smallness of knowledge of the fine arts possessed by the latter "; and a passing remark made by Prof. Huxley, in one of his essays, would seem to imply a disposition to draw the line of separation between animal and human intelligence at this point. Prantl regards the phrase " die Kunsttriebe der Thiere " as a metaphorical expression involving a confusion of terms, since animals, with all their apparent artistic ability and taste shown in constructing and decorating their habitations, do not seek to embody ideas in material forms—an assumption which begs the very question in dispute. Schiller, in his well-known poem, Die Künstler, makes man's pre-eminence consist solely in his artistic faculty:

> In Fleiss kann dich die Biene meistern,
>   In der Geschicklichkeit ein Wurm dein Lehrer sein,
> Dein Wissen theilest du mit vorgezogenen Geistern,
>   Die Kunst, o Mensch, hast du allein.
>
> In diligence the bee can master thee,
> In skilfulness a worm thy teacher be,
> Knowledge thou dost with higher spirits own,
> But art, O man, thou dost possess alone.

Herbart, as we have already seen, does not recognise this demarcation. "If one asks for a specific characteristic of mankind, which is not physical, but spiritual, original and universal, and does not resolve itself into a more or less, I confess," he says, "that I do not know of any such distinction and do not think it exists." He then enumerates the advantages possessed by man—namely, hands, speech, and a long and helpless infancy, to the use and influence of which are due the extraordinary growth of the human brain in size and complexity and the corresponding development of intellectual power. In the acuteness of his senses and in many peculiarities of physical structure man is inferior to some of the lower animals. He has not, says Prof. Cope, kept pace with other mammals in the development of his teeth, which are "thoroughly primitive"; his nose is less serviceable than that of the dog; the eagle has a far better eye; the ankle joint of the sheep is, as a piece of mechanism, stronger and less liable to derangement than the corresponding joint in man; the horse's foot consists of a single compact elastic toe, on which the animal runs while its heel is carried in the air and never touches the ground, thus attaining a springiness and swiftness of motion beyond the reach of the human plantigrade. Whatever lightness and elasticity of step man possesses is due less to the perfection of his bodily organism than to the uplifting influence of his intellect. With the decay of his mental powers *Homo sapiens* slouches like a bear, as may be observed in the ungainly and unsteady gait of cretins and idiots, however vigorous they may be physically.

The objection urged by Prof. Kedny against the doctrine of evolution—namely, that man's helpless infancy proves him to be different in kind from other animals—ignores the fact that the soko and many other species of the genus *Simia* pass through a period of infant helplessness almost as long as that of some savage tribes. The babyhood of the anthropoid apes is much longer and more helpless than that of the cynopithecoids, the platyrhines, or the lemurs; and the higher the order of the monkeys, the more they resemble man in this respect. Mr. Wallace captured a young orang-outang, which had to be fed and cared for like a human infant, lay rolling on the ground with all fours in the air, and could hardly walk when it was three months old; whereas a macacus of the same age seemed to have already acquired full use of its limbs and mental faculties. The long duration of this complete dependence on parental care in the case of the human infant, so far from disproving the doctrine of evolution, furnishes one of the strongest arguments in its favour, since it helps to explain how man gradually attained his intellectual primacy among the primates. The American platyrhines, marmosets, and other smaller long-tailed monkeys reach maturity in three or four years, whereas the African dog-headed apes require ten or twelve years for their full development, and with the larger anthropoids this period of growth is nearly as long as with human beings.

The fact that quadrumans have flexible organs of prehension, can grasp and handle things and imitate human actions, gives them a great advantage over quadrupeds. A dog may be as intelligent as a chimpanzee, but he is unable to "show off" as well; he can not untie knots with his paws, nor put on clothes, nor eat with knife and fork, nor uncork bottles, nor drink wine by lifting the glass to his lips, nor use a toothpick, nor perform a variety of tricks which make the monkey appear to be relatively far more richly endowed with mental gifts than is actually the case, and throw into the shade the most conspicuous exploits of the poodle and the collie.

Nevertheless, this manual and digital dexterity can scarcely be overestimated as a means of disciplining the mind and increasing the volume of the brain; and if chimpanzees, orang-outangs, and sokos had enjoyed the thousands of years of domestication and thorough breeding and training, from which dogs have so immensely profited, there is no knowing what advances in knowledge and acquisitions of intellectual culture they might not have made. It is wonderful how much they learn through observation and very slight instruction during a few months' intercourse with human beings, discharging with evident pleasure the duties of body servant or waiter, answering the door bell, showing visitors into the parlour, fetching water, kindling the fire, washing dishes, turning the spit, and doing all sort of chores in and about the house. " Such an ape," says Brehm, " one can not treat as a beast, but must associate with as a man. Notwithstanding all the peculiarities it exhibits, it reveals in its nature and conduct so very much that is human, that one quite forgets the animal. Its body is that of a brute, but its intelligence is almost on a level with that of a common boor. It is absurd to attribute the actions of such a creature to unthinking imitation; it imitates to be sure, but as a child imitates an adult, with understanding and judgment."

That the plastic and progressive period of the monkey's individual development is short, and that its faculties become set and stationary at a comparatively early age, is undeniable; but the same holds true of the negro, who loses his educability and ceases his mental growth much earlier than the Caucasian. The longer or shorter duration of this formative season in the mental life of man is, to some extent, a matter of race, but in a still greater degree the result of civilization.

The hand is also a valuable instrument for the cultivation of the æsthetic sense, and the more flexible and sensitive this instrument becomes, the greater are the results achieved by it in this direction. But there are animals without hands that show an appreciation of the beautiful. Mr. Darwin has proved conclusively that birds take pleasure in sweet sounds and in brilliant colours, and that the sentiment thus awakened and appealed to plays an important part in the preservation and perfection of the species through natural selection. The struggle for existence is not always carried on by fierce combat and the triumph of brute force, but quite as frequently takes the form of competition in beauty, addressing itself either to the ear as alluring song or to the eye as attractive plumage; and the bird that possesses these characteristics in the highest degree carries off the prize in the tournament of love, and propagates its kind.

There is no doubt that birds take delight in the gorgeousness of their own feathers, and the more brilliant their hues the greater the vanity they display. Conspicuous examples of this love of admiration and fondness of parading their finery are the peacock and the bird of paradise.

The decoration of its boudoir by the bower bird, as described by Mr. Gould in his History of the Birds of New South Wales, indicates a decided and discriminative preference for bright and variegated objects, and evinces no small amount of æsthetic feeling and artistic taste in selecting and arranging them. The bower is built of sticks and slender twigs gracefully interwoven, so that the tapering points meet at the top, and is adorned with the rose-coloured tail feathers of the inca cockatoo and the gay plumes of other parrots, tinted shells, bleached bones, rags of divers hues, and whatever gaudy or glittering trinkets may please the bird's fancy. Sometimes the space in front of the bower is covered with half a bushel of things of this sort, laid out like a parterre with winding walks, in which the happy possessor of the garnered treasures struts about with the pride and pleasure of a connoisseur in a gallery of paintings, or a bibliophile who has his shelves filled with *incunabula* and other rare editions. These objects have often been brought from a great distance, and are of no possible use to the bird except as they gratify its love of the beautiful and appeal to what we call in man the æsthetic sense. Its conduct can be explained in no other way; for the bower is not a nest in which eggs are laid and hatched and young ones reared; it is a *salon* or place of social entertainment, and thus serves a distinctly ideal purpose.

A similar artistic talent is shown by the African and Asiatic varieties of weaver, which suspend their nests from the slender branches of trees over running water and thus render them inaccessible to monkeys and other plundering foes; sometimes, too, they weave into them long thorns with the points turned outwards, so that their house becomes a castle and resembles a fortress bristling with bayonets. It is also a significant fact that the nests of young birds are loosely and clumsily built and not constructed to perfection until the third year, proving that their skill is a gradual acquirement, something learned, to a certain extent, by practice and instruction, and not purely instinctive. Among Western representatives of this class of bird artists the Baltimore oriole, the European titmouse (*Parus pendulinus*), and the Brazilian japu are the most noteworthy.

The singing of birds, as a means of sexual attraction, implies a certain appreciation of melody. Indeed, many of them do not confine themselves to the songs of their species, but learn notes from other birds and snatches of tunes from musical instruments. Canaries can be taught a variety of airs by playing them repeatedly on a piano or on a hurdy-gurdy. They listen with attention and imitate the strains which take their fancy. If harmony or the concord of sweet sounds, as distinguished from melody or the simple succession of sweet sounds, does not enter into bird music, the same may be said of the music of primitive man and of all early nations. Savages, like feathered songsters, sing in unison, but not in accord.

There are also some remarkable instances of the musical education of unmusical birds, so that they learn songs wholly foreign to their species. A recent case of this kind occurred at the little town of Tannendorf, in the principality of Reuss, in Germany. It is well known that the sparrow has naturally no gift of song, but keeps up a tedious and often intolerable chirping; at the same time it is by no means a stupid bird. An invalid soldier of Tannendorf, named Pfeifer, succeeded in training one of these birds into a very superior songster, which took the first prize for its vocal powers at the exhibition of the Ornithological Society "Ornis," held at Leipsic in February, 1896. Isaak Walton says the nightingale " breathes such sweet, loud music out of her little instrumental throat, that it might make mankind to think that miracles are not ceased." But the miracle is still greater when such " sweet descants " come from the throat of the sparrow, whose natural notes are so shrill and monotonous. This incident is a striking proof of the musical capabilities of the unmusical passerine family, and shows what wonderful results may be attained by the patient development of the faculties of the lower animals. We may add that one of the jurors who awarded the prize to Pfeifer's sparrow was Prof. Göring, of Gera, well known for his scientific explorations in Brazil.

Spiders, locusts, and lizards show a decided love of musical tones whether produced by themselves for the purpose of sexual attraction or by the human voice and performers on instruments. It is related of Beethoven that when in his boyhood he was learning the violin in his room, a spider used to let itself down from the ceiling on the instrument and remain there so long as he kept on playing. One day his mother entered the room and seeing the spider in its accustomed place obeyed the instincts of a careful housewife and killed it, whereupon the youthful Ludwig, angry at this brutal treatment of his silent and attentive auditor, flung his fiddle on the floor and smashed it. Beethoven was once questioned as to the truth of this statement and declared that he had no recollection of any such incident, but, on the contrary, had good reason to believe that every living creature, including flies and spiders, would have got as far as possible away from his horrid gratings on the catgut. The tale in this case may be a fiction, but authentic instances of this kind have occurred and been recorded by musicians.

It is possible, however, that the apparent fondness of spiders for music and their supposed partiality for the tones of stringed instruments may be due to the resemblance of such noises to the buzzing of flies when caught in a web. This explanation, if it be correct, would throw a blur upon the evidence adduced in proof of the æsthetic endowments of the *Arachnida*, and show that they are "not moved with concord of sweet sounds," but rather by their eagerness for prey, and are therefore "fit for treason, stratagems, and spoils." Thus, for example, during a concert in the celebrated Gewandhaus at Leipsic, Prof. Reclam watched the movements of a spider, which let itself down from a chandelier when the violin solos were performed, but hurried back and disappeared as soon as the full orchestra began to play. The illusion of the buzzing fly was suddenly dissipated by the clangour of trumpets, the rattle of drums, the clash of cymbals, and the deep notes of the ophicleide and trombone.

It is true, however, that certain species of spiders produce musical tones and seem to take pleasure in them. This is the case with the gigantic spider of Central Australia, known to zoölogists as *Phrictis crassipes*. According to recent observations made by Prof. Baldwin Spencer, it is from six to seven centimetres long and measures twelve centimetres between the extremities of its legs. By rubbing its feelers against a comblike set of bristles on the back part of its body, it brings forth sounds that in a still night may be heard for a distance of two or three metres.

In some charming verses entitled The Lark (Die Lerche) a Westphalian poetess, Annette von Droste-Hülfshoff, gives a vivid description of a spring morning on a North German heath and the orchestral performance, in which the cricket plays the kit, the beetle the horn, the gnat the triangle, and the bumblebee the bass viol:

> So tausendstimmig stieg noch nie ein Chor,
> Wie's musicirt aus grünem Heid hervor.

The Greeks ascribed to the cicada (τέττιξ) a beautiful voice (φωνή), which seems to have been to their ear the synonym and supreme ideal of melodiousness. Plato calls this insect the prophet of the Muses, and Anacreon extols it as the divinest of singers. Only the males were supposed to be endowed with this fine vocal gift, hence the witty suggestion of Xenarchos that men might well envy the happiness of the Cicadæ, whose females are dumb (ὧν ταῖς γυναιξὶν οὐδ' ὁτιοῦν φωνῆς ἔνι). As a matter of fact, the cicada has no voice at all, but is a superior violinist, and always carries with him his fiddle organically attached to his person. The recent investigations of the entomologists Vitus Graber and Brunner von Wattenwyl have first given a clear idea of the construction of this musical apparatus and the manner of its use. The arrangement and the mode of operation differ somewhat in different species. Perhaps the purest violin tones are produced by the *Locusta cantans,* popularly known in some parts of Europe as the "harvest bird." The male is an unwearied wielder of the fiddle stick, and the female will listen for hours with evident rapture to his performances. She, too, possesses rudimentary organs of the same kind, but they are visible only under the microscope and not sufficiently developed to produce tones. Clearly her musical education has been neglected and she has not made the most of the gifts with which Nature has endowed her; but as a good listener she is unrivalled, and finds ample opportunity to cultivate this rare and amiable talent.

Not only do some species of monkeys, like the chimpanzees and sokos, get up concerts of their own in the depths of the forest, but dogs, which are generally supposed to be decidedly unmusical, also discriminate between tunes and express their preferences or aversions in an unmistakable manner. A friend of mine, who had a magnificent St. Bernard dog, was fond of playing the violoncello. The dog used to lie quietly in the room with closed eyes, and appeared to pay no attention to the music until his master struck up a certain tune, when the dog immediately and invariably sat up on his haunches and began to howl. If the tune which called forth such emotions had been written on a very high key, or characterized by shrill tones or harsh dissonances, the conduct of the dog might be easily explained. But such was not the case. There was nothing in this piece more than in any other, so far as any one could observe, that ought to grate the canine ear. Many incidents of this kind might be cited to prove that even dogs are not indifferent to musical compositions, and show a nice discrimination between them, having their likes and dislikes, as well as human beings.

Indeed, the howling of a dog under such circumstances is no proof that the sounds are painful to him; on the contrary, it is probably his manner of expressing his appreciation of them. The noise he makes may be disagreeable to us, just as his sharp bark often is, but this is no reason why it should not be an utterance of joy. More than half a century ago, the zoöpsychologist Scheitlin suggested that the dog takes pleasure in the musical tones, and merely wishes to accompany the performer. If they were so discordant as to be distressful to him, he could leave the room; but he has never been known to seek relief in this manner. It is also certain that he imitates in some degree what he hears, and that the howls stimulated by the lengthened notes of the organ differ from those excited by the piano, the violin, or the human voice. The Rev. A. Treiber, a clergyman in Richen near Eppingen, Germany, states that when a student in the university he had a female poodle named Rolla, who was very fond of singing with him. Thus, for example, if he began to sing the Lorelei, especially in falsetto, Rolla would strike in, and one could easily perceive how she would try to catch the tune by following, though not very successfully, the ascending and descending notes of the melody. Still more striking instances of this kind are given by Alix (L'Esprit de nos Bêtes, p. 364 sq.), of several poodles that sang the scale perfectly, and one that "sang very agreeably a magnificent piece by Mozart." This remarkable dog belonged to Habeneck, the director of the Paris opera, and not only had the superior advantage of living in a musical atmosphere, but had also received special musical instruction.* We know that the imitative impulse in dogs, and more particularly in poodles, is very strong, and there is no reason why it should not extend to the imitation of articulate and musical tones, so far as the structure of the vocal organs render their reproduction possible. That there is also an element of æsthetic gratification in such performances would seem to be evident from the fact that some tones are imitated in preference to others.

"Mammals and birds," says a recent writer, "possess the habit of indulging frequently in more or less regular or set performances, with or without sound, or composed exclusively of sound; and these performances, which in many animals are only discordant cries and choruses, and uncouth, irregular motions, in the more aërial, graceful, and melodious kinds take immeasurably higher, more complex, and more beautiful forms. . . . We see that the inferior animals, when the conditions of life are favourable, are subject to periodical fits of gladness, affecting them powerfully and standing out in vivid contrast to their ordinary temper. And we know what this feeling is—this periodic intense elation which even civilized man occasionally experiences when in perfect health, and more especially when young. There are moments

---

* Cf. Karl Groos, Die Spiele der Thiere, p. 183.

when he is mad with joy, when he can not keep still, when his impulse is to sing and shout aloud and laugh at nothing, to run and leap and exert himself in some extravagant way. Among the heavier mammalians the feeling is manifested in loud noises, bellowings, and screamings, and in lumbering, uncouth motions—throwing up of heels, pretended panics, and ponderous mock battles. In smaller and livelier animals, with greater celerity and certitude in their motions, the feeling shows itself in more regular and often more complex ways. . . . Birds are more subject to this universal joyous instinct than mammals, and there are times when some species are constantly overflowing with it; and as they are so much freer than mammals, more buoyant and more graceful in action, more loquacious, and have voices so much finer, their gladness shows itself in a greater variety of ways with more regular and beautiful motions, and with melody." * Here we have very near approaches to conscious artistic production. Indeed, the theory of the derivation of the æsthetic sentiments from the play impulse, enunciated by Schiller more than a century ago in his Briefe über die ästhetische Erziehung des Menschen, and more systematically formulated by Herbert Spencer in his Principles of Psychology, would make these sentiments common both to men and animals. The play impulse, like the art impulse, has its source in the imagination, and the form in which it finds expression is determined chiefly by the force of heredity modified by the action of the imitative instinct. The kitten, the kid, the puppy, the young bird, and the child has each

---

* The Naturalist in La Plata, by W. H. Hudson, 3d ed., London, 1895, pp. 264, 280.

its own style of performance, which consists in what Herbert Spencer calls "dramatizing" the serious and habitual occupations of their parents and elders. The young do in fun the things in which the adults of their kind are earnestly engaged, and thus unconsciously exercise faculties and acquire capabilities destined to be of immense ulterior benefit to them in the struggle for existence. This form of diversion, which Prof. Groos calls "experimenting," enters largely into the education of children, and is the most important factor in kindergarten instruction, owing to the long duration of human infancy. With most animals this sportive simulation begins very early, but soon gives place to serious activity. The kitten stretches its legs, thrusts out its claws, scratches whatever comes in contact with them, runs after a rolling ball, and, in lack of other objects of pursuit, indulges the predatory instincts of the feline race by chasing its own tail. The lowest organisms have no leisure in the proper sense of the term; all the powers they possess are constantly and exclusively employed in securing the bare necessities of life. As we ascend to animals of a higher type we find them endowed with superior faculties, which enable them to procure food and shelter, to protect themselves against enemies, and to propagate their species, and yet leave them time and strength not wholly absorbed in making provision for their pressing wants. This surplus of energy finds its natural outlet in play, which, as already observed, is the resultant of hereditary tendencies and imitative propensities, and marks the starting point in the cultivation of the ideal. In the above-cited work (p. 227), Mr. Hudson describes very vividly the strange impression produced by large flocks of chakars or crested screamers singing together on the pampas of South America. On one occasion he saw countless numbers of them gathered along the shores of a narrow sheet of water and arranged in several well-defined groups of about five hundred each and extending all round the lake. "Presently one flock near me began singing and continued their powerful chant for three or four minutes; when they ceased, the next flock took up the strains, and after it the next, and so on until the notes of the flocks on the opposite shore came floating strong and clear across the water—then passed away, growing fainter and fainter, until once more the sound approached me travelling round to my side again. The effect was very curious, and I was astonished at the orderly way with which each flock waited its turn to sing, instead of a general outburst taking place after the first flock had given the signal." At another time he heard a similar performance on a still larger scale. This occurred at a place called Gualicho, on the southern pampas, where, after riding over a marshy plain covered with innumerable groups of chakars, he had stopped for the night at a small *rancho* inhabited by a *gaucho* and his family. About nine o'clock, while they were eating supper, the vast multitude of birds covering the marsh for miles around burst forth into a tremendous evening song, the effect of which was indescribable. "One peculiarity was that in this mighty noise, which sounded louder than the sea thundering on a rocky coast, I seemed to be able to distinguish hundreds, even thousands, of individual voices. Forgetting my supper, I sat motionless and overcome with astonishment, while the air and even the frail *rancho* seemed to be trembling in that tempest of sound. When it ceased, my host remarked with a smile,

'We are accustomed to this, señor—every evening we have this concert.' It was a concert well worth riding a hundred miles to hear." It is evident from the regular occurrence of this performance and the circumstances attending it that it was merely a musical entertainment, having nothing more to do with sexual solicitation or wooing than has an assembly of men and women giving or hearing a piece of instrumental or vocal music in a theatre or a concert hall. It was the expression and gratification of æsthetic feeling, having a crude and inchoate artistic character like the singing of savages, only more melodious. As just stated, it is well known that monkeys, and especially chimpanzees and gorillas, take a childish delight in making loud and discordant noises by beating on hollow trees and other resonant objects, and often accompanying this din with shouts of exultation, which afford them the same pleasure that it gives an urchin to pound on a tin pan or many an adult to listen to the monotonous wheezing of a hand-organ. The only difference in these cases is that the bird has a finer musical sense and a more delicate appreciation of the "concord of sweet sounds" than the simian or the human creature. The fact, too, that some birds sing less freely and to our ear at least less charmingly in the pairing season than at other times, would imply the existence of other and stronger incentives to song than sexual attraction, and can be best explained by assuming that they find pleasure in the mere act of singing or in the production of musical tones. For this reason they meet together and exercise their voices in concert; this occurs also as a general rule out of the pairing season and after the young are fledged and can take part in the performances. It has also been repeatedly observed that the male sings his most beautiful songs not as a suitor, but as the prospective father of a family, namely, while the female is brooding. Whatever emotion this more elaborate carol of the bird may express, whether paternal pride or conjugal love, it is certainly not the strain in which the feathered warbler woos his mate. Some species of grallatorial birds, such as the New Caledonian kagu (*Rhinochitus jubatus*) and the African umber or shadow-bird (*Scopus umbretta*), although extremely grave and dignified when in repose, are wont to work off their surplus of vigor by wild pranks and antic postures. All at once the usually staid and rather ungainly fowl begins to dance and skip about in the liveliest manner, seizing with its long beak a tail feather or the tip of its wing, as a ballet-dancer does her gauzy skirt with the tips of her fingers, and prancing and pirouetting in a style that would do credit to any Terpsichorean "star" of the operatic stage. Sometimes the fantastic performance ends with a startling acrobatic climax, the bird standing on its head, or rather on the end of its beak, flapping its wings and waving its bright-hued legs like flames in the air.

The fertilization and propagation of many plants depend upon the existence of a sense of colour in insects, and the exercise of choice in the selection of flowers. This preference implies a pleasure in certain hues, and consequently the possession of a rudimentary perception of beauty. Plants whose fecundation depends upon the action of the wind do not develop such a variety of colours as those in which this depends upon the agency of insects. Nature can trust her ill-favoured daughters to the wooing of the wind, but if she wishes to attract a nicer class of suitors she must endow her children with brilliant qualities.

The power of distinguishing between colours has been denied not only to the lower animals, but also to the lower races of mankind. But a more extended and accurate knowledge shows that the conclusion is incorrect in both cases. We know that the American aborigines discriminate between the seven primary colours, and it is absurd to infer that this faculty was wanting to the Homeric men merely because we do not find all these colours mentioned in the Homeric poems. It has also been asserted that the ancient Assyrians could not distinguish green from blue or yellow, because no word was found for it in the remains of their language. But the tiles discovered at Nineveh prove that they had a very clear conception and æsthetic appreciation of the distinction between yellow, green, and blue, and probably did not confound any colours of the solar spectrum. The evidence of language on this point is purely negative and necessarily defective.

Even the religious sentiment, which has been assumed to be the peculiar possession of man, is fairly foreshadowed in the lower animals. The unanimity of opinion among those who have made the most careful study of this subject, and whose views are therefore entitled to the greatest consideration, is quite remarkable. M. A. de Quatrefages, in his Rapport sur le Progrès de l'Anthropologie (Paris, 1867, p. 85), maintains that "domestic animals are religious, since they readily obey those who appeal to them with the rod or with sugar." In other words, they are amenable to rewards and punishments, doing the will and seeking to win the favour of superior beings, on whom they are dependent, propitiating and fawning upon them, creeping and grovelling on the ground in abject adoration, in order to assuage their anger or to secure their kind regard. "There is no difference," adds the same author, "between the negro who worships a dangerous animal, and the dog who crouches at his master's feet to obtain pardon for a fault. . . . Animals fly to man for protection as a believer does to his god."

This is precisely the feeling of the savage in respect to the superior skill and power of the civilized man. *Taguta kipini te Atua*—doctor all the same as God—are the words in which the Morioris, or aborigines of the Chatham Islands, expressed their sense of dependence on a higher agency, whose beneficent workings they perceived but could not comprehend. Among rude tribes the sentiment of devotion to a chief does not differ essentially from that of devotion to a god; the Romans, at the height of their civilization, paid divine honours to their emperors; and in modern monarchies kings are officially addressed in terms of reverential awe and superlative adulation as all-wise and all-powerful beings, whose favour one can not sufficiently implore with servile words and suppliant knee.

"The feeling of religious devotion," says Darwin, "is a highly complex one, consisting of love, complete submission to an exalted and mysterious superior, a strong sense of dependence, fear, reverence, gratitude, hope for the future, and perhaps other elements. No being could experience so complex an emotion until advanced in his intellectual and moral faculties to at least a moderately high level. Nevertheless, we see some distinct approach to this state of mind in the deep love of a dog for his master, associated with complete submission, some fear, and perhaps other feelings." *

---

* The Descent of Man. London, 1874, p. 95.

Comte held that the higher animals are capable of forming fetichistic conceptions, and of being strongly influenced by them. Herbert Spencer denies the truth of this statement in its absolute form, because it does not fit into his theory of the origin and evolution of religious ideas, but admits, what is essentially the same thing so far as the present discussion is concerned, that "the behaviour of intelligent animals elucidates the genesis" of fetichism, and gives two illustrations of it. "One of these actions was that of a formidable beast, half mastiff, half bloodhound, belonging to friends of mine. While playing with a walking stick, which had been given to him and which he had seized by the lower end, it happened that in his gambols he thrust the handle against the ground, the result being that the end he had in his mouth was forced against his palate. Giving a yelp, he dropped the stick, rushed to some distance from it, and betrayed a consternation which was particularly laughable in so large and ferocious-looking a creature. Only after cautious approaches and much hesitation was he induced again to lay hold of the stick. This behaviour showed very clearly that the stick, while displaying none but the properties he was familiar with, was not regarded by him as an active agent, but that when it suddenly inflicted a pain in a way never before experienced from an inanimate object, he was led for the moment to class it with animate objects, and to regard it as capable of again doing him injury. Similarly, in the mind of the primitive man, knowing scarcely more of natural causation than a dog, the anomalous behaviour of an object previously classed as inanimate suggests animation. The idea of voluntary action is made nascent, and there arises a tendency to regard the object with alarm, lest it should act in some other unexpected and perhaps mischievous way. The vague notion of animation thus aroused will obviously become a more definite notion as fast as the development of the ghost theory furnishes a specific agency to which the anomalous behaviour can be ascribed."

This conduct of the dog, which every one must have observed under similar circumstances, corresponds to that of the savage who worshipped an anchor which had been cast ashore, and on which he had hurt himself when he first came in contact with it. Superstitious fear of this sort prevails most among men of the lowest order of intelligence, or in that stage of society in which human beings are psychically least removed from beasts. In proportion as they rise in the scale of existence and unfold their mental faculties, the more they free themselves from the tyranny of the supernatural. The terror of the dog hurt by the stick was out of all proportion to the pain inflicted, and arose solely from the fact that it was produced by a mysterious cause; it was fear intensified by the intervention of a ghostly element, and thus working upon the imagination it assumed the nature of religious awe. The case is analogous to that of a big, burly, brutal savage trembling before a rude stock or stone, or a Neapolitan bandit cowering before an image of the Virgin or kissing devoutly the feet of a crucifix.

The other illustration given by Herbert Spencer is that of a retriever, who, associating the fetching of game with the pleasure of the person to whom she brought it, would often fetch various objects and lay them at her master's feet; and "this had become in her mind an act of propitiation."

THE RELIGIOUS SENTIMENT. 355

Still more interesting and instructive are Mr. Romanes's experiments with a Skye terrier. This dog, which was exceedingly intelligent and therefore an excellent subject for psychological study, " used to play with dry bones, by tossing them in the air, throwing them to a distance, and generally giving them the appearance of animation, in order to give himself the ideal pleasure of worrying them. On one occasion, therefore, I tied a long and fine thread to a dry bone and gave him the latter to play with. After he had tossed it about for a short time I took the opportunity, when it had fallen at a distance from him and while he was following it up, of gently drawing it away from him by means of the long, invisible thread. Instantly his whole demeanour changed. The bone which he had previously pretended to be alive, began to look as if it were really alive, and his astonishment knew no bounds. He first approached it with nervous caution, but, as the slow receding motion continued and he became quite certain that the movement could not be accounted for by any residuum of force which he had himself communicated, his astonishment developed into dread, and he ran to conceal himself under some articles of furniture, there to behold at a distance the 'uncanny' spectacle of a dry bone coming to life." In this instance we have the exercise of close observation, judgment, reason, and imagination culminating in the exhibition of superstitious fear—all the elements, in short, which constitute religious sentiment in its crudest form.

Animals are afraid of darkness for the same reason that children are. Thunder, lightning, and other violent meteorological phenomena, which inspire the primitive man with awe and therefore play a prominent part in the evolution of early mythology, produce a similar impression upon many of the lower animals, simply because they are mysterious noises which appeal to the imagination and stimulate the mythopœic faculty. Mr. Romanes states that "on one occasion, when a number of apples were being shot out of bags upon the wooden floor of an apple room, the sound in the house as each bag was shot closely resembled that of distant thunder." A setter was greatly alarmed at the noise until he was taken to the apple room and shown the cause of it, after which "his dread entirely left him, and on again returning to the house he listened to the rumbling with all cheerfulness." Dogs and horses can be completely cured of their fear of thunder by being present at artillery practice; they imagine that they now know what produces the dreadful roar, and are henceforth free from all apprehension concerning it.

To some extent this sense of the supernatural seems to enter into the sphere of pure imagination and to excite in the minds of animals those vague feelings of anxiety and alarm arising from mere figments of the brain and characterized as superstition. The following incident, "illustrating the instinctive fear of death and consciousness of its presence manifested by birds," is related by Buist: "A hen canary died, was buried, the nesting establishment broken up, the surviving cock bird removed to a new cage, and the hatching cage itself thoroughly cleansed and purified, and put aside till the following spring. Never, however, could any bird afterward endure being placed in that cage. They fought and struggled to get out, and, if all in vain their efforts, they moped, huddling close together, thoroughly unhappy, refusing to be comforted by any amount of sunshine, companionship, or dainty food." The experiment was tried with foreign birds, that had not been in the house when the death of the hen occurred, and could not, therefore, have known anything of the melancholy event by observation. The result, however, was always the same. " For the future that cage to them was haunted."

It is a common belief that many animals can see ghosts and future events. Justinus Kerner declares (Die Seherin von Prevost, i, 125) that they are endowed with second sight, and that numerous facts can be adduced in proof of it. This uncanny faculty is supposed to be especially strong in dogs and horses. Storks, too, are known to have foreseen the burning of houses on which they had been wont to build their nests, and to have abandoned them, taking up their abode on other buildings or on trees in the vicinity. No sooner had the anticipated conflagration taken place, and a new house been erected on the same site, than they returned and built their nests on it as heretofore. That Balaam's ass perceived the angel, which was beyond the ken of the prophet, ought to suffice to convince every believer in the plenary inspiration of the Bible of the spectre-seeing powers of the lower animals. The ghost stories told of dogs and horses are quite as numerous and well authenticated as those which have been told of men. There is no psychological theory of apparitions that does not explain these strange phenomena as satisfactorily in beasts as in human beings. The night side of Nature casts its gloom over both.

Of course, if religion is a direct and special revelation to man, then no sentient creature prior and inferior to him could have any share in it. The hypothesis of a pure primitive monotheism, of which all polytheistic systems of belief are mere distortions and degradations, would also tend to exclude the lower animals from the possession of religious sentiment by showing that the religious history of the race has been a downward instead of an upward movement, a corruption instead of an evolution. Its growth would not correspond to the growth of intelligence, and it could no longer be studied as a psychological phenomenon, but would be removed at once from the province of scientific investigation. There can be no science of the supernatural, since science recognises only the operation of natural laws. A miracle that can be explained, as the rationalistic school of theology has attempted to do, ceases thereby to be a miracle. The essence of religion is mystery; the sole aim of science is to clear up and thus do away with mysteries—a goal which it is always tending toward but will never reach, for the same reason that an asymptotic line never meets the curve which it is constantly approaching.

BIBLIOGRAPHY.

Alix, E. L'Esprit de nos Bêtes. Paris, 1890.

Altum, Bernard. Der Vogel und sein Leben. 5th ed. Münster, 1875.

Aimé, Bernard. De l'Aphasie et de ses diverses Formes. Paris, 1885.

Anima Brutorum secundum sanioris philosophiæ Canones Vindicata. Naples, 1742.

Austin, Philip. Our Duty toward Animals. London, 1885. A melancholy survival of mediævalism.

Autenrieth, J. H. F. von. Ansichten über Natur und des Seelenlebens. Stuttgart, 1836.

Bain, A. The Senses and the Intellect. 3d ed. London, 1868.

Baldwin, J. M. Handbook of Psychology. Senses and Intellect. 2d ed. New York, 1890.

Bateman. F. On Aphasia or Loss of Speech. London, 1870.

Bechstein, J. M. Naturgeschichte der Stubenvögel. 4th ed. by Lehmann. Halle, 1840.

Bentham, Jeremy. Introduction to the Principles of Morals and Legislation. London, 1789.

Bingley, William. Animal Biography. 4 vols. 6th ed. London, 1824.

Birds' Courts of Justice, in Popular Science Monthly. New York, Oct., 1888.

Boccardo, G. L'animale e l'uomo. Preface to vol. vii of Biblioteca dell' Economista, iii ser. Torino, 1882.

Bouillaud, Jean. Recherches cliniques propres à démontrer que le sens du langage articulé et le principe coordinateur des mouvements de la parole résident dans les lobules antérieures du cerveau. Paris, 1848.

Bregenzer, Ignaz. Thier-Ethik. Darstellung der sittlichen und rechtlichen Beziehungen zwischen Mensch und Thier. Bamberg, 1894.

Brehm, A. E. Thierleben. Allgemeine Kunde des Thierreichs. With numerous illustrations and colored plates. 10 vols. 2d ed., Leipzig, 1884; 3d ed., 1892. New edition thoroughly revised with a supplemental volume by Pechuël-Loesche, 1890–'95.

Brehm, C. L. Beitrage zur Vogelkunde. Neustadt, 1821, *sqq*.

Broca, P. Sur le Siége de la Faculté du Langage Articulé.

Büchner, Ludwig. Aus dem Geistesleben der Thiere oder Staaten und Thaten der Kleinen. 4th revised ed. Leipzig, 1895. [Devoted especially to the intellectual faculties of ants, bees, and spiders.]

Büchner, Ludwig. Liebe und Liebesleben in der Thierwelt. 2d ed. Leipzig, 1885.

Budde, E. Naturgeschichtliche Plaudereien. Berlin, 1891.

Carus, K. G. Vergleichende Psychologie. Wien, 1866.

Carus, K. G. Psyche. zur Entwickelungsgeschichte der Seele. 2d ed. Stuttgart, 1851.

Caspari, Otto. Die Urgeschichte der Menschheit mit Rücksicht auf die natürliche Entwickelung des frühesten Geisteslebens. 2 vols. Leipzig, 1873.

Chambre de la. Traité de la Connoissance des Animaux où tout ce qui a esté dict pour et contre le raisonnement des est examiné. Paris, 1648.

Conn, H. W. Evolution of To-day. New York, 1886.

Cope, E. D. Origin of the Fittest. London, 1887.

Cornish, C. J. Animals at Work and Play. London, 1896.

Darwin, Charles. On the Origin of Species. London, 1859.

Darwin, Charles. The Descent of Man. London, 1872–'79.

Darwin, Charles. Expression of the Emotions in Men and Animals. London, 1872.

Darwin, Charles. Variation of Animals under Domestication. 2 vols. London, 1885.

Dax, G. L'Aphasie. Paris, 1878.

Delamétherie, I. C. De l'Homme considéré moralement.

Dilly, A. D. Traité de l'Ame des Bestes. Lyon, 1676.

Ebrard, E. Études de Mœurs. Genève, 1863.

Eimer, G. H. Th. Die Entstehung der Arten. Jena, 1888.

Espinas, Alfred. Des Sociétés Animales, Étude de Psychologie Comparée. 2d ed. Paris, 1878.
Fischer, L. Über das Prinzip der Organisation und die Pflanzenseele. Mainz, 1883.
Fletcher, Ralph. A Few Notes on Cruelty to Animals. London, 1846.
Flourens, P. De l'Instinct et de l'Intelligence des Animaux. 3d ed. Paris, 1851.
Flügel, O. Das Seelenleben der Thiere. 2. verm. Aufl. Langensalza, 1886.
Flügel, O. Über die Instincte der Thiere mit besonderer Rücksicht auf Romanes und Spencer. Zeitschrift für exacte Philosophie. Bd. xvii. 1890.
Forster, T. Philozoia or Moral Reflections on the Actual Condition of the Animal Kingdom, etc. Brussels, 1839.
Foveau de Courmelles, F. V. Les Facultés Mentales des Animaux. Paris, 1890.
Fuchs, C. J. Das Seelenleben der Thiere im Vergleich mit dem Seelenleben der Menschen. Erlangen, 1854.
Galton, Francis. Inquiries into Human Faculty and its Development. London, 1883.
Garner, R. L. The Speech of Monkeys. New York, 1892.
Garner, R. L. Gorillas and Chimpanzees. London, 1896.
Garratt, G. Marvels and Mysteries of Instinct. London, 1862.
Gerlach. Die Seelenthätigkeit der Thiere. Berlin, 1859. Cf. Magazin für Thierheilkunde, Jahrg. xxv. Heft 2.
Gompertz, Lewis. Moral Inquiries on the Situation of Man and Brutes. London, 1824.
Graber, Vitus. Die Insekten. 2 vols. München, 1879. Vols. xxi and xxii of the scientific library Die Naturkräft.
Groos, Karl. Die Spiele der Thiere. Jena, 1896.
Guer, M. Histoire Critique de l'Âme des Bêtes. 2 vols. Amsterdam, 1749.
Haeckel, Ernst. Über die Entstehung und den Stammbaum des Menschengeschlechts. 4th ed. Berlin, 1881.
Hagen, H. Über die Lebensweise der Termiten und ihre Verbreitung. 1852.
Hahn, Eduard. Die Hausthiere und ihre Beziehungen zur Wirthschaft des Menschen. Leipzig, 1896.
Helps, Arthur. Some Talk about Animals and their Masters. London, 1873. Chatty and kindly.

Hennings, J. C. Geschichte von den Seelen der Menschen und Thiere. Jena, 1744.

Herder, J. G. Ideen zur Philosophie der Geschichte der Menschheit. 4 vols. Riga, 1791. See especially his Briefe zur Beförderung der Humanität."

Hertwig, O. Die Symbiose oder das Genossenschaftsleben im Thierreiche. Jena, 1883.

Hildrop, John. Free Thoughts upon the Brute Creation. London, 1742.

Hippel, Robert von. Die Thierquälerei in der Gesetzgebung des In- und Auslands, historisch, dogmatisch und kritisch dargestellt, nebst Vorschlägen zur Abänderung des Reichsgesetzes. Berlin, 1891.

Hoffmann, L. Thierpsychologie. Stuttgart, 1881.

Houssay, F. Les Industries des Animaux. Paris, 1889.

Houzeau, J. C. Étude sur les Faculté Mentales des Animaux comparées à celles de l'Homme. 2 vols. Mons, 1872.

Huber, François. Nouvelles Observations sur les Abeilles. Genève, 1792.

Huber, Pierre. Recherches sur les Mœurs des Fourmis Indigènes. Paris et Genève, 1810.

Hudson, W. H. The Naturalist in La Plata. 3d ed. London, 1895.

Jackson, T. Our Dumb Companions. London, 1889.

Janet, P. L'Automatisme Psychologique. 2 vols. Paris, 1894.

Joannet, l'Abbe Claude. Les Bêtes mieux connues ou le pour et contre l'Âme des Bêtes. 2 vols. Paris, 1770. A defense of Descartes against Bouiller's Essai sur l'Âmes des Bêtes.

Joly, Henri. Psychologie Comparée. L'Homme et l'Animal. Paris, 1877.

Kind, Aug. Teleologie und Naturalismus in altchristliches Zeit. Jena, 1875.

Kipling, J. Lockwood. Man and Beast in India. London, 1893.

Kirchner, Fr. Über die Thierseele. Halle, 1890.

Krause, K. C. F. Abriss des Systems der Philosophie des Rechts. Göttingen, 1828.

Kussmaul, Adolf. Untersuchungen über das Seelenleben des neugeborenen Menschen. 2d ed. Tübingen, 1884.

Kussmaul, Adolf. Die Störungen der Sprache. 3d ed. Leipzig, 1885. Published in Ziemssen's Handbuch der speciellen Pathologie und Therapie.

Laboulaye, Honoré. Essai de Psychologie Morale ; les Animaux et l'Homme. Paris, 1878.
Landois, H. Thierstimmen. Freiburg im B., 1874.
Lawrence, John. A Philosophical Treatise on Horses, and on the Moral Duties of Man towards the Brute Creation. 2 vols. London, 1796–'98.
Legroux, Dr. De l'Aphasie. Paris, 1875.
Lenz, H. O. Gemeinnützige Naturgeschichte. 3d ed. Gotha, 1851.
Leroy, Ch. G. Lettres Philosophiques sur l'Intelligence et la Perfectibilité des Animaux. 2d ed. Paris, 1802. Published originally in 1764 as the work of a "physicien de Nuremberg" in order to evade persecution by the Sorbonne. German translation, Nürnberg, 1807.
Lindsay, W. L. Mind in Lower Animals in Health and Disease. 2 vols. London, 1879.
Lindsay, W. L. Physiology and Pathology of Mind. London, 1871.
Lubbock, J. On the Senses and Intelligence of Animals, with Special Reference to Insects. London, 1889.
Maisonneuve, P. L'Instinct et l'Intelligence. Paris, 1884.
Marshall, W. Die Papageien. Leipzig, 1889.
Marshall, W. Leben und Treiben der Ameisen. Leipzig, 1889.
McCook, H. C. The Natural History of the Agricultural Ants of Texas. Philadelphia, 1879.
McCook, H. C. The Honey Ants of the Garden of the Gods, and the Occident Ants of the American Plains. Philadelphia, 1882. Cf. Proceedings of the Academy of Natural Sciences of Philadelphia for 1878.
Meier, G. F. Versuch eines neuen Lehrgebäudes von den Seelen der Thiere. Halle, 1750.
Menault, E. L'Intelligence des Animaux. Paris, 1872.
Mivart, St. George. The Cat. London, 1881.
Moggridge, J. T. Harvesting Ants and Trapdoor Spiders. London, 1873.
Mohnike, O. G. J. Affe und Urmensch. Münster, 1888.
Montanari, A. Trattenimento Metafisico intorno ai Principali Sistemi dell' Anima delle Bestie, etc. Verona, 1761.
Morgan, C. Lloyd. Animal Life and Intelligence. London, 1891.
Morgan, C. Lloyd. Comparative Psychology. London, 1894.
Morgan, C. Lloyd. Habitat and Instinct. London, 1897.

Morgan, L. H. The American Beaver and his Works. Philadelphia, 1868.

Morris, F. O. Animal Sagacity and Character. London, 1861.

Morris, F. O. Anecdotes in Natural History. London, 1888.

Müller, A. and K. Wohnungen, Leben und Eigenthümlichkeiten in der höheren Thierwelt. Leipzig, 1869.

Müller, A. aud K. Thiere der Heimath. Cassel, 1881.

Müller, Max. The Science of Language. 2 vols. London, 1891. Cf. a criticism of this work entitled Max Müller and the Science of Language, by William Dwight Whitney. New York, 1892.

Naumann, J. F. Naturgeschichte der Vögel Deutschlands. 12 vols. Leipzig, 1822-'44. Remarkable repository of observations made by his father, J. A. Naumann, and himself. Continued by Blasius, Baldamers, and Sturm, 1852, *sqq.*

Neumann, Karl. Liebe, Ehe und Eheleben in der Vogelwelt. Sammlung gemeinverstänlich. wissenschaftl. Vorträge. Heft 169. Berlin, 1893.

Nicholson, Edward Byron. The Rights of an Animal, a new Essay in Ethics. London, 1879.

Nicholson, George. On the Conduct of Man to Inferior Animals. Manchester, 1797.

Noiré, Ludwig. Der Ursprung der Sprache. Mainz, 1877.

Orphal, W. C. O. Der Philosoph im Walde. Leipzig, 1807.

Orphal, W. C. O. Sind die Thiere sinnliche Geschöpfe ? Leipzig, 1811.

Pereira, Gomez. Antonia Margarita opus nempe physicis, medicis ac theologis non minus utile quam necessarium. Anno M.D.LII. An elaborate work of pp. 831, fol., with an appendix containing objections by Prof. Michael Palacios and Pereira's reply.

Pertz, Max. Über das Seelenleben der Thiere. Leipzig, 1876.

Peters, Wilhelm. Über das Wohnen und Wandern der Thiere. Berlin, 1867.

Pierquin de Gembloux, C. C. Idiomologie des Animaux ou Recherches historiques, anatomiques, physiologiques et glossologiques sur le Langage des Bêtes. Paris, 1844.

Pierquin de Gembloux, C. C. Traité de la Folie des Animaux et de ses rapports avec celle de l'homme et les législations actuelles. Paris, 1839.

Porphyrius. De Abstinentia ab esu Animalium. Venet, 1547.
Preyer, W. Die Seele des Kindes. 3d ed. Leipzig, 1890.
Primatt, Humphrey. A Dissertation on the Duty of Mercy and Sin of Cruelty to Brute Animals. London, 1776.
Quatrefages, A. de. Rapport sur le Progrès de l'Anthropologie. Paris, 1867.
Radau, R. Die Lehre vom Schall. München, 1869.
Reclam, K. H. Geist und Körper in ihren Wechselbeziehungen. Leipzig, 1859.
Reimarus, H. S. Allgemeine Betrachtungen über die Triebe der Thiere, hauptsächlich über ihre Kunsttriebe. 1st ed. Hamburg, 1762; 4th ed., 1798.
Ribot, Th. L'Hérédité Psychologique. 5th ed. Paris, 1894.
Rokitansky, Karl Freih. von. Die Solidarität alles Thierlebens. Wien, 1869.
Rolph, W. H. Biologische Probleme, zugleich als Versuch einer rationellen Ethik. Leipzig, 1882.
Romanes, G. J. Animal Intelligence. London, 1881.
Romanes, G. J. Mental Evolution in Animals with Essay on Instinct. London, 1883.
Russ, K. Die sprechenden Papageien. 2d ed. Magdeburg, 1887.
Russ, K. Allerlei sprechendes Gefiedertes Volk. Magdeburg, 1889.
Salt, Henry S. Animals' Rights, with an essay on Vivisection in America, by Albert Leffingwell, M. D. New York, 1894.
Sbaragli, J. H. Entelechia seu anima sensitiva Brutorum demonstrata contra Cartesium. Bonn, 1716.
Schaller, J. Das Spiel und die Spiele. Weimar, 1861.
Scheitlin, P. Versuch einer vollständigen Thierkunde. Stuttgart und Tübingen, 1840.
Schelling, F. W. Über die Kunsttriebe der Thiere. System der Naturphilosophie. Jena, 1799.
Schmarda, Ludwig Karl. Andeutung aus dem Seelenleben der Thiere. Wien, 1846.
Schneider, G. H. Der thierische Wille. Leipzig, 1880.
Schneider, G. H. Der menschliche Wille. Berlin, 1882.
Schopenhauer, Arthur. Parerga und Paralipomena. Berlin, 1851.
Schultze, Fritz. Die Thierseele. Leipzig, 1868.
Schultze, Fritz. Die Sprache des Kindes. Leipzig, 1880.
Schultze, Dr. Fritz. Vergleichende Seelenkunde. Leipzig, 1897.

Selenka, E. Studien über die Entwickelungsgeschichte der Thiere. Wiesbaden, 1883.

Semper, Karl. Die natürlichen Existenzbedingungen der Thiere. 2 vols. Leipzig, 1880.

Steinthal, H. Einleitung in die Psychologie und Sprachwissenschaft. Berlin, 1871.

Stiebeling, J. C. Über den Instinkt des Huhns und der Ente. New York, 1872.

Stiebeling, J. C. Naturwissenschaft gegen Philosophie. New York, 1871.

Strümpell, Ludwig. Die Geisteskräfte der Menschen vergleichen mit denen der Thiere. Leipzig, 1878.

Strümpel, Ludwig. Vorschule der Ethik. Mitau, 1845.

Traité de l'Âme des Bêtes avec des Reflexions Physiques et Morales. Paris, 1737.

Tissot, J. Psychologie Comparée. De l'Intelligence et de l'Instinct dans l'Homme et dans l'Animaux. Paris, 1879.

Trögel, F. M. Causeries sur la Psychologie des Animaux. Leipzig, 1856.

Tschudi, F. von. Das Thierleben der Alpenwelt. 11th ed. Leipzig, 1890.

Virey, J. J. Histoire des Mœurs et de l'Instinct des Animaux. 2 vols. Paris, 1822.

Vogt, Carl. Altes und Neues aus Thier- und Menschenleben. 2 Bde. Frankfort, 1859. Comprises the substance of two former works: Untersuchungen über Thierstaaten and Bilder aus dem Thierleben.

Voltaire, F. M. Arouet de. Dictionnaire Philosophique. Art. Bêtes.

Waitz, Theodor. Die Anthropologie der Naturvölker. 5 vols. Leipzig, 1860-'67.

Wasmann, Erich, S. J. Instinct und Intelligenz im Thierreich. Ein kritischer Beitrag zur modernen Thierpsychologie. Freiburg im Breisgau, 1897. An attempt to revive the doctrines of Thomas Aquinas and the mediæval scholastics.

Wasmann, Erich, S. J. Vergleichende Studien über das Seelenleben der Ameisen und der höheren Thiere. Freiburg im Breisgau, 1897. Denies intelligence to ants, and denounces Brehm, Büchner, and Lubbock as "superficial zoöpsychologists."

Watson, J. S. Reasoning Power in Animals. London, 1867.
Weinland, F. Eine Vogelfamilie. Der Zoologische Garten. 1861.
Weismann, A. Gedanken über Musik bei Thieren und Menschen. Deutsche Rundschau lxi (1889).
Wenzel, G. I. Neue auf Vernunft und Erfahrung gegründete Entdeckungen über die Sprache der Thiere. Wien, 1800.
Whitney, William Dwight. Language and the Study of Language. New York, 1867.
Whitney, William Dwight. Oriental and Linguistic Studies. New York, 1872.
Willis, Th. De Anima Brutorum quæ hominis vitalis ac sensitiva est. Amstelod, 1674.
Wilson, Andrew. Sketches of Animal Life and Habits. London, 1877.
W. de F. (Antoine Edmond Wolheim). Animaux Diplomates. Leipzig, 1863.
Wood, J. G. Man and Beast, Here and Hereafter. London, 1874.
Wundt, Wilhelm. Vorlesungen über die Menschen- und Thierseele. 2 vols. Leipzig, 1863–'64. 2d ed. 1 vol. 1892. English translation. London, 1896.
Youatt, William. The Obligation and Extent of Humanity to Brutes, etc. London, 1839.
Young, Thomas. An Essay on Humanity to Animals. London, 1798.
Ziegler, H. E. Über den Begriff des Instincts. Verh. der Deutschen Zoolog. Gesellschaft. Leipzig, 1891.

INDEX.

Abbott, Dr., on mechanical skill of the oriole, 200; on aphasia in the quail, 299.
Abel, his offering, 90.
Addis, W. E., on animals as things, 97.
Æschylus, quoted, 31.
Agriculture, conservative influence of, 39; first promoters of, 43; holiness of, 60–69; fatal to nomadic life, 71.
Ahuramazda, 59, 62, 66, 67.
Airyana-Vaêjô, 64.
Akemmanô, 69.
Albert, meaning of, 22.
Alexander, meaning of, 22.
Aliens, as enemies, 26; former treatment in England, France, and Italy, 34–37.
Alix, E., on musical poodles, 345.
Altum, on the "natural necessity" of bird-song, 220.
Amenophis III and IV, 33.
Americans, their extravagance, and ignorance of economic laws, 102, 103.
Amnon, his relations to Tamar, 51.
Amphicyon, ancestor of dog and bear, 281.
Amtsberg, his experiment with a pike, 186.

Anacreon, extols the song of the cicadæ, 342.
Anasis, foes of agriculture, 72.
Anaximander, as an evolutionist, 137.
Ancon sheep, origin of, 214.
Angrô-Mainyush, 69.
Animals, denial of their rights by mediæval and modern schoolmen, 2, 96–99; in the eyes of primitive man, 4; superstitious fear and worship of, 6, 118–120; effect of domestication on their relations to man, 7; Zarathustra, doctrine of, 8, 59; regard of Buddhists and Brahmans for, 9, 88, 148, 164; Greek speculation concerning, 10; views of early Christians concerning, 11; moral and religious symbolism of, 12; capital punishment and excommunication of, 13, 155; penal laws for the protection of, 14, 100–102; the clergy opposed to rights of, 14; Lotze's theory of animal souls, 15; measure of man's duty to, 18; anthropocentric teachings of Holy Writ, 10, 88–90; practical workings of biblical teachings concerning, 89–93, 96–99, 146–152, 156–161; arguments in favour of

immortality of, 94; limitation of man's dominion over, 97; cruelties inflicted in transporting, 101; influence of the doctrine of metempsychosis, 135-138; societies for protection of, 138; hospitals for, 139-144, 147; excessive scruples of Jainas concerning, 141; tormented by Mantegazza, 142; foolish fondness for pet, 145; Munich Thierschutzverein, 149; dilemma of a philozoic parson, 149; scriptural injunctions of kindness to, 149-152, 164; in hagiology, 152-159; "not Christians," 160; attitude of Catholic Church toward, 159-162; decree of Pius IX concerning, 160; cruelty of Italians to, 154, 160; inalienable rights of, 164; psychical kinship with man, 167, 171; power of choice in the lowest, 172; no precise line of demarcation between vegetables and, 171-173; Schneider's psychological classification of, 175, 178-181, 185; Oken's temperaments of, 179; mental impulses of food-storing, 179; sentinel-posting, 182, 183; mutual benefit associations of, 184; didactic procedures with, 186-189; paternal training of, 189, 198; wild speculations about, 191; fanciful distinctions between man and, 192-194; difference in bodily constitution and its influence on mental development of man and, 194-196; institutions common to man and, 197; communities of, 198; improvableness of, 198, 200-206, 212-216; military organization of, 210; inheritable qualities of, 213-215; their conceptual world compared with that of savages, 215, 225; influence of domestication on their mental development, 216-218; tamability of, 218; "time-sense" in, 223-227; tradition in communities of, 225; suicide of, 227; nature of benevolence in, 228; conjugal unions of, 228, 229; jealousies of, 229; sense of community in, 230; courts of justice held by, 230-235; criminal impulses in, 236; social and industrial organizations of, 227-240; adaptation to environment, 240-243; relapse into barbarism, 241; different stages of evolution, 247; grades of intelligence, 250; size of brain and mental capacity. 252; personal benevolence and altruism of, 255-257; use of tools by, 257-265; as miners, 263; logical faculty of, 265-267; humour in, 268; speech as the Rubicon between man and, 271-273, 291-295, 300; formation of general concepts by, 282, 286, 291, 310; ability to count, 285; strange transformations of individual, 280; evolution of different species of, 281; vices of civilization acquired by, 313; language of, 291-294, 299-310, 310-330; knowledge of the fine arts as a distinction between man and, 333; inferiority of man in physical structure to, 334, 335; long and helpless infancy of some, 335; advantage of quadruman over quadruped, 336; æsthetic sense and artistic skill of, 337-339; appreciation of melody by, 339-350; musical instruction of, 340, 344, 345; play impulse in, 345-347; musical apparatus of, 343; musical performances by, 343-350; colour sense in, 87, 350; religious senti-

INDEX.

ment in, 351-358; fear of thunder and darkness, 355, 356; fear of ghosts and second sight in, 356, 357.

Anthony, St., his sermon to fishes, 155; celebration of his feast in Rome, 156.

Anti-Semitism, a survival of tribalism, 50.

Ants, progressive evolution of, 205; structure of their hills, 206; white, 207; agricultural, 245-247; cazadores or nomadic, 247; "cattle-lifting," 248; slaveholding, 249; difference of intelligence in, 250; ingenuity of, 251; size of brain and mental capacity of, 252; babyhood and education of, 253; moral attributes of honey, 255; Darwin's experiment with, 255; method of heating their habitations, 265; language of gesture highly developed by, 292; language of, 311, 318.

Apes, social organization of, 41; wine-making and pottery-fabricating, 261; language of, 315, 317, 324; long infancy of anthropoid, 335; manual dexterity of, 262, 336. See MONKEYS.

Aphasia, cause and examples of, 295-298.

Aphides, kept by ants as cattle, 248, 249.

Apiarists, improvements introduced by, 202.

Apuleius, his Golden Ass, 115; reputation as a sorcerer, 116.

Aquinas. See THOMAS AQUINAS.

Archæopteryx, ancestor of birds and reptiles, 281.

Ardà-Vîrâf, his piety and incest, 13.

Aristophanes, quoted, 23.

Aristotle, quoted, 10; on the re-embodiment of poets as cygnets, 114; on heart-beating as peculiar to man, 191; on phôné and psophos, 309.

Armaiti, personification of the earth, 59, 66.

Arnold, Thomas, perplexed by the Christian doctrine of animals, 91, 92, 93.

Arnold, T., quoted, 97.

Arts, first developed by dwarfs and cripples, 44; animal appreciation of the fine, 333-351.

Asceticism repudiated by Parsis, 61.

Ashemaogho, the impure, 63.

Asia, meaning of, 22.

Asperena, meaning of, 62.

Ass, use of weapons by an, 259; vision of Balaam's, 357.

Assyrians, their distinction of colours, 351.

Astôvîdhôtus, the death demon, 63.

Astrology, persistence of, 85.

Astronomy, geocentric, 24, 82.

Augustine, on sorcery, 115; on predestination, 130.

Austin, Philip, denies the duty of kindness to brutes, 97.

Australians, as beasts of venery, 78; their numerals and words for colours, 287.

Avesta, quoted, 2, 51, 59-63.

Bacon, Lord, quoted, 93.

Bactria, social evolution in, 58.

"Bai Sakarbai," hospital for animals, 142.

Balantis, tribal ethics of the, 25.

Barbarian, origin of the word, 23; later applications of the term, 31, 32; conservatism of the, 70.

Barclay, John, on the immortality of animals, 94.

Barley, first grain cultivated, 44.

Barnum, his contract with King Thibo for a white elephant, 143.

372 INDEX.

Barthelot, Major, his cruelty to Africans, 78.
Bastian, on termites in India, 211.
Bateman, on aphasia, 298.
Battas, parricide as a mark of filial affection, 109.
Bayle, on comets as portents, 86.
Bedouins, hospitality of, 30; hostility to agriculture, 72.
Bees, colonization of, 198; artificial comb for, 202; products of human industry used by, 203; "sweating," 203–205; cause of the hexagonal form of their cells, 205; liability to error, 207, note; mathematical thinking of, 224; democratic government of, 237; maternal functions and conjugal relations of their queen, 237–239; drones as prince consorts, 239; radical changes in the habits of, 240; degeneracy of, 240–242; brigand, 241; effect of alcoholic drinks on, 242; interhival relations of, 243; foresight of, 244; brain of, 252; pantomimic language of, 292.
Beethoven, his musical spider, 341.
Belief, supersession of kinship by religious, 53. See RELIGION.
Bell, A. Graham, his attempts to teach animals to speak, 216, 300.
Bellarmine, Cardinal, his nasty means of grace, 152.
Benares, as the centre of the earth, 21.
Bentham, Jeremy, on animal rights, 13.
Bergh, Henry, on the cruelty of corporations to animals, 101, 102.
Berlin, meaning of, 22.
Bernard of Clairvaux, on the toleration of Jews, 74.
Bernardine de Saint-Pierre, his teleology, 83.

Berzelius, on the products of vital forces, 174.
Bethlehem, as the centre of the earth, 21.
Bettzieh-Beta, on termites, 208.
Bible, anthropocentric teachings of the, 149, 150, 152.
Birds, Jewish protection of, 151; Italian cruelty to, 154; relation of saints to, 154, 157; sentinel-posting, 182; conjugal instinct in, 197; parasitic, 199; their improvements in nest building, 200–202; "time-sense" in, 226; conjugal virtue of, 229; sense of community in, 230; courts of justice held by, 230–234; use of tools by, 260; their ability to count, 285; aphasia in, 299; articulation by, 300 (see PARROTS and RAVENS); æsthetic sense in, 338, 339; musical education of, 329, 340; the play impulse in, 345–350; musical concerts by, 347–349; dancing, 350.
Blanchard, on termites, 208.
Blood, brotherhood of, 24–26; superstitious regard for, 26–28; Christian enlargement of the bond of, 28, 29; all nations and all creatures of one, 164. See ETHICS.
Body, development of the soul dependent upon the, 15; seat of the soul, 26; made by the soul, 121; mind not destroyed with the, 123; acquisition of new organs of the, 124; no knowledge of spirit separate from the, 132–135.
Boëthius, on metamorphoses, 116.
Bordeaux, termites in, 211.
Bouillard, on aphasia, 295.
Bower bird, æsthetic sense and artistic taste of the, 338.
Bozen, ingenuity of ants in, 251.

INDEX. 373

Brahmanism, doctrine concerning animals, 9, 88, 138; caste in, 58; necessity of a son to salvation, 63.
Brehm, A. E., on the language of apes, 315; on the nearness of the ape to man, 337.
Broca, on aphasia, 295, 297.
Brotherhood, the sole cement of primitive society, 25; Christian theory of universal, 28; Cicero and Marcus Aurelius on human, 28, 29; artificial creation of, 27; feigned by sovereigns, 33; natural superseded by religious, 55.
Büchner, quoted, 137; on drones, 238; on demoralized bees, 242; on the rearing and training of ants, 253, 254.
Buddha and Buddhism, doctrine of animal life, 9, 88, 127, 136, 138.
Buffon, on animal intelligence, 202, 212, 213, 291.
Buist, on canaries and the haunted cage, 356.
Bullfights, under the auspices of the Church, 161.
Burnaburiash, his correspondence, 33.
Burns, quoted, 90, 131.
Butler, Bishop, on animal immortality, 94.

Cain, type of primitive man, 30; his offerings, 90.
Calvin, John, his doctrine of predestination, 130, 168.
Canaries, conjugal virtue of, 229; appreciation of melody by, 339; ghost-seeing, 356.
Cannibalism, origin of, 27, 118; practised by Europeans, 79.
Canning, George, his abolition of the alien law, 38.
Carlyle, Thomas, his false etymology of king, 41.

Carrion fly, acts under the impulse of perception, 180.
Carus, Paul, on naming, 291; on the language of ants, 318.
Cassiodorus, quoted, 191.
Cathrein, Victor, ridicules kindness to animals, 99.
Cato, on usury, 74.
Cats, asylum for, 144; training of, 219; standard of goodness in, 228; the play impulse in, 347.
Cattle, care for, 2; cruelty in transporting, 101, 102.
Cazadores, nomadic ants, 247.
Cells of bees, cause of their form, 205.
Celsus, his polemic against anthropocentric Christianity, 10.
Cicero, cosmopolitan spirit of, 28.
Chakar, 64.
Chakars, musical concerts of, 348, 349.
Chaumette, his aviary, 145.
Chekiang, wine-making apes in, 261.
Chester, meaning of, 22.
Chevage, tax on aliens, 36.
Chimpanzees. See APES and MONKEYS.
China, the centre of the earth, 23.
Chinvad, bridge to paradise, 63, 132.
Christ, his gospel a sword, 56.
Christianity, its attitude toward animals, 10, 88-99, 188.
Cicada, prized by the Greeks as a singer, 342; now known to be a violinist, 343.
Clifford, Prof., on the movements of atoms, 202.
Cobbe, Frances Power, on zoöphily, 3.
Cock, cruel treatment of the, 160, 161.
Cockneyism, a survival of tribalism, 38.

Colour, sense of, in animals and savages, 287, 338, 351.
Columbanus, St., wild animals attracted by, 153.
Comets, as portents, 86.
Comte, Aug., on fetichism in animals, 353.
Conception, impulse of, 175–181.
Congo, cruelties in, 76.
Consanguinity, the primitive basis of moral obligation and social union, 25. See BLOOD and BROTHERHOOD.
Consciousness, in the lowest forms of life, 167.
Cope, Prof., on superiorities of animals to man in physical structure, 334.
Cormorants, tool-using, 260.
Couthon, his pet spaniel, 145.
Cowper, William, quoted, 151, 193.
Creed, cohesive attraction of the, 56.
Cripples, the first inventors, 44.
Crows, tool-using, 260.
Crystallization, phenomena of, 172–174.
Cuckoo, habits of different species of, 199.
Cyril, quoted, 39.
Cyrus, wedded to his sister, 51.
Czynski, Czeslav, hypnotic suggestion and swindling, 117.

Dadabhoi Naoroji, a "nigger," 46.
Dakotas, as dog-eaters, 118.
Dards, primitive barbarism of, 70.
Darmesteter, his solar theory of Zarathustra, 69.
Darwin, Charles, on intelligence in the oyster, 17; his Origin of Species, 14, 137; on worms, 151; on the fallibility of bees, 207, note; on agricultural ants, 245; on the ant's brain, 252; his experiment with ants, 255; on the different barking tones of the dog, 282; on the sense of melody and colour in birds, 337; on religious sentiment in animals, 352.
Darwin, Erasmus, his theory of divine beneficence, 103.
Dasyus, aborigines of India, 68.
Death, continuous process of, 184.
Descartes, on animals as machines, 170.
Detractus personalis, punishment for emigration, 40.
Devas, diabolized deities, 59.
Dogs, tortured by Mantegazza, 142; asylum for, 144; as specialists, 216; reared for food, 218; "time-sense" in, 226; generous friendship of, 256, 257; ak-ak denoting eagerness in the language of, 282; distinct barking tones acquired through domestication, 282, 283; power of classification, 286; ability to count, 288, 322; expression by wagging the tail, 292; at the mercy of metaphysicians, 294; able to articulate words, 300; sent on errands and to the market, 321–323; compared to monkeys, 336; musical taste and training of, 343–345; religious sentiment in, 352–355; ghost stories about, 357.
Domestication, results of, 216–219, 282, 283.
Donkey, club wielding, 259.
Droit d'aubaine, against aliens, 35.
Drones, the prince consorts of the hive, 239.
Droste-Hülshoff, Annette von, her poem The Lark, 342.
Dybowski, on Garner's studies of simian speech in Africa, 330–332.

INDEX. 375

Ebrard, on ants, 254.
Edinburgh, meaning of, 22.
Edmonson, Dr., on the judicial proceedings of hooded crows, 231.
Eleusis, privilege of asses at, 150, 151.
Elliott, George, quoted, 90.
Emerson, Ralph Waldo, on evolution, 11, 185.
Emigration, right of, 47–49; barbarizing effects of, 78–80.
Empedocles, 137.
English, insularity of the, 23; boor, 34; laws against aliens, 37, 38; tribal spirit in their extradition treaties, 40; King George's men, 49; treatment of Negritos, 78; cannibalism, 79.
Épaves, aliens as waifs, 36.
Ethics, relation of animal psychology to evolutional, 2, 4, 14; philozoic, 3; ethnocentric, 24–26, 73; survivals of tribal, 33–40, 49, 50, 77–79; supersession of ethnocentric by theocentric, 53, 55–57, 58, 73–76; anthropocentric, 89; defect of Jewish and Christian, 88–99; scientific basis of philozoic, 96; legal recognition of animal rights as a corollary to evolutional, 100.
Europe, meaning of, 22.
Expatriation, right of, 47.

Family, included domestic animals, 8; differentiated out of the tribe, 41, 229.
Fate, Oriental idea of, 121, 122.
Feigning death, by animals, 181.
Fellahin, strangers hated by, 34; meaning of, 72.
Feuerbach, Ludwig, 137.
Fire, used by animals, 264.
Fiske, John, Herbart's idea developed by, 194.

Flourens, on mind in animals, 291.
Flügel, O., on animal tribunals and simulation, 234, 235.
Forel, on ants, 254.
Fournier, his fondness for squirrels, 145.
Francis of Assisi, St., wolf of Gubbio reformed by, 155; his Cantico delle Creature and sympathy with beasts and birds, 157, 158.
Frederick III, his decree concerning Jews in Nuremberg, 74.
Free will, nature and extent of, 167–170.
French, childish exhibition of tribalism by the, 25.
Fuligatti, biography of Cardinal Bellarmine, 152.

Gabella hereditarius, against aliens, 40.
Galen, on man's upward look, 192.
Gall, St., bear addressed in Latin by, 153.
Galton, Francis, on training animals for mental power, 215.
Garner, R. L., his studies of simian speech and their results, 312–315, 319, 328–332.
Gatha dialect, 58.
Gennadius, on creation, 84.
Geography, ethnocentric, 20–24, 82.
George, meaning of, 22.
Germans, called Němici, 24; ethics of ancient, 25; called barbarians by the Romans, 32; old laws of the, 35, 36.
Gervinus, on the ancient Greeks' love of nature, 190.
Gessner, K., on the stargazer, 184.
Gifford, Ellen M., her refuge for cats and dogs, 144.
Glendower, 85, 153.
Goar, St., hinds commanded by, 153.

Goethe, horoscopy of, 86; on preexistence with Frau von Stein, 116; on Orient and Occident, 162; on language, 289.
Göring, Prof., 340.
Gould, on the bower bird, 338.
Gruber, Vitus, on the relative size of insect brains, 252; on the musical apparatus of the cicada, 343.
Graul, on the treatment of animals in India, 136.
Gray, quoted, 87.
Greeks, early speculations of the, 14; survival of their geographical terms, 22; their classification of mankind, 23; their attitude to strangers, 31; their disregard for the decrepit, 139.
Green, Prof., on the mental operations of a burnt dog in the presence of fire, 294.
Grenville, Lord, his alien bill, 38.
Groos, Karl, on the play impulse in animals, 347.

Habeneck, his musical poodle, 345.
Haeckel, Ernst, on treatment of animals in India, 136; on aboriginal ants, 205.
Hagen, H., on termites, 209, 211.
Hallam, Henry, on the souls of brutes, 95.
Hamann, 137.
Hand, its influence on mental growth, 194, 195.
Hanikl, his remarkable parrot, 326.
Hartmann, Eduard von, 137.
Harvey, anticipated by Nemesius, 11.
Haug, 65.
Hawthorne, his Donatello, 159.
Hebrews, ethnocentric notions of, 6; survival of tribal marriage among, 8; tribal religion of, 57;
contempt for Christians, 75. See JEWS.
Hedley, John Cuthbert, animal rights denied by, 96.
Hegel, quoted, 55.
Heidegger, Gotthard, his speaking raven, 300.
Heine, Heinrich, reflections of his lizard at Lucca, 189.
Hens, effect of alcohol on, 243.
Hephæstus, 44.
Heraclitus, on psophos, 309, 310.
Herbart, on the disciplinary value of religion in early society, 55; on the three chief causes of man's mental development, 194, 334.
Herbert, George, his precepts for parsons, 90.
Herder, on animals as elder brothers, 18, 137.
Hermit crab, 184.
Herodotus, quoted, 43, 57, 63; on the transmigration of souls, 110.
Hettinger, his apology for Christianity, 14.
Hickok, Dr., denies duties to animals, 98.
Hippocrates, on the stargazer, 110.
Hobbes, man rational because orational, 272.
Holden, Prof., on the suicide of a rattlesnake, 227.
Homer, on kindness to strangers, 29.
Horace, his apokyknosis, 114.
Horoscopy, 85.
Horses, cruelty to, 102; training of, 188; transmission of qualities by, 213; "time-sense" in, 226; fear of thunder, 356; ghost-seeing, 357.
Hospitality, sacredness of, 30; tokens of, 32.
Hospitals for animals, 139–145, 147.
Hotspur, his retort to Glendower, 85.

INDEX. 377

Huber, François, on bees, 207, note; "royal treatment" of, 237; on foresight in, 244.

Huber, Pierre, on ants, 207; on "royal treatment" of ants, 237; on slaves of ants, 254.

Hubert, St., his vision of a stag, 155.

Hudson, W. H., on the play impulse in mammals and birds, 345, 346; on the musical concerts of chakars, 347-349.

Humboldt, A. von, on an Aturian parrot, 325.

Humboldt, W. von, on language, 289.

Hun, Dr., on child language, 277.

Hunger, as an impulse, 55, 175-177.

Huxley, his line of separation between animals and man, 383.

Hyænarctos, ursine and canine, 281.

Impulses, Schneider's four categories of, 175.

Incarnation, as a curse, 113.

Infusoria, their power of choice, 170; metamorphosed into algæ, 172.

Insects, sense of colour in, 87, 350; vocal organs of, 310-312; love of musical tones, 340; musical instruments of, 342, 343.

Instincts, analogous to habits in man, 18; not unchangeable, 198, 210, 212, 240-243; liable to error, 206, 207.

Institutions common to animals and man, 197.

International conscience, slow growth of, 25.

Ionic school, evolution taught by the, 9, 137.

Ivan the Terrible, 100.

Jaeger, Charles, on the language of animals, 294.

Jainas, excessive fear of killing animals, 136, 140, 141.

James, St., care for birds, 154.

Jameson, Mr., his barbarity to negroes, 78.

Jameson, Mrs., on the delinquencies of the pulpit, 90; on cruelty to dogs in Vienna, 145.

Janet, Charles, on the vocal organs of ants, 312.

Japu, artistic talent of the, 339.

Jerusalem, the centre of the earth, 21.

Jesse, on beehive fortifications, 241.

Jesus, his belief in pre-existence, 114.

Jews, their ethnocentric ethics, 23, 24; superstitious regard for blood, 26; not proselytizing, 57; tribal feeling, 74-76; their schabbesgoi and dread of perjury, 76; place of animals in their cosmogony and religion, 88-91; theory of transmigration adopted by, 110-112; poverty of their scriptures in regard to animals' rights, 149-152.

Julian, Emperor, compares German to the caw of ravens, 293.

Jus albinagii, kolbekerlii, and wildfangiatus, 35, 36.

Kagu, dancing feats of the, 350.

Kalewala, quoted, 4.

Kapila, Indian prototype of Darwin, 166, note.

Kedney, Prof., his argument against evolution, 335.

Kerner, Justinus, on second sight in animals, 357.

Kepler, 85.

Kilima-Njaro, a raid of driver ants in, 247.

Kinship. See BLOOD and BROTHERHOOD.

Kipling, J. Lockwood, on the treatment of animals in India, 135.
Kitto, on the sacrificial institution, 90.
Knownothingism, 39.
Kolbenrecht, 36.
Krause, on mental development in relation to moral rights, 18, 137.

Lactantius, Lucius, on human brotherhood, 28, 29; his distinction between men and brutes, 192.
La Mettrie, on men as machines, 170.
Landois, Prof., on the tone language of mosquitoes, 310, 311.
Language, as a barrier between man and beast, 271-273; as the test of rationality, 272; roots of, 273-276, 278, 281, 283, 302, 303, 315; child, 276-280; identity of thought and, 288, 317; a social institution, 288, 289; theories of the origin of, 278, 283, 290, 306-308; not supernatural, 306; pantomimic, 289, 291, 292, 299; organ of articulate, 295; animal, see ANIMALS.
La Rochelle, termites in, 209, 211.
Lecky, on Egyptian hospitals for animals, 145.
Leibnitz, on the use of human speech by animals, 216, 300.
Lenz, on the maternal training of kittens, 188; on tradition in bee-hives, 243.
Leonhard, St., feast of, 157.
Lessing, on the acquisition of new senses, 124-126.
Levant, local meaning of, 22.
Liberia, republic of, 77.
Lindo, Miss, her hospital for horses, 144.
Linsecom, Dr., studies of Texan agricultural ants, 245, 246.
Livingstone, on the soko, 267-269.

Lizard, Heine's, 189; love of music, 340.
Lotze, Hermann, his theory of souls, 15, 137.
Louis XIV, absolutism of, 100.
Lubbock, Sir John, his experiment with ants, 250.
Lucian, his imaginary metamorphosis, 115.
Lucretius, on human brotherhood, 29.
Ludwig I, his futile efforts to attract birds, 159.

Macarius, St., his penance, 153.
Macgowan, on wine-making apes, 261.
Magpie, its ability to count four, 285.
Maine, Sir Henry Sumner, on the evolution of social institutions, 25-40, 45, 46, 53.
Man, his place in nature, 15, 83-89, 99; turning point in his evolution, 17; world of primitive, 20; ethics of primitive, 24, 73; brotherhood of, 28, 33; social survivals of palæozoic, 33, 34-38; supremacy of, 89-99, 192-194; his superiority and inferiority in bodily structure, 194, 334.
Manchuria, wine-making apes in, 261.
Mandrill, vicious nature of the, 312-315; teachableness of the, 313.
Manichæans, metempsychosis taught by the, 112.
Mansel, Prof., on language and thought, 288.
Mantegazza, Prof., hospitals for animals ridiculed by, 142; his "tormentor," 142.
Manu, institutes of, 121.
Manyuema, their inferiority to the soko, 267, 268.

Marat, his devotion to doves, 145.
Marcus Aurelius, his cosmopolitan spirit, 29.
Marriage, tribal, 31; with next of kin, 51, 52; with ritual relations prohibited, 54; among animals, 197, 228.
Mather, Increase, on portents, 86.
Mazdayasnians, 59, 69.
McCook, Henry C., on agricultural ants, 247; on honey ants, 255.
Mecca, centre of the earth, 21.
Mechanical labour, its value as a discipline, 195.
Megapode, artificial heat used by the, 265.
Melody, appreciated by birds, insects, and dogs, 339-349; monkeys' attempts to produce, 269, 349.
Menander, on the brotherhood of man, 28.
Mercer, G. R., on the language of the Veddahs, 290.
Mersenne, on the language of animals, 323, 325.
Metempsychosis, 9, 88; universality of the belief in, 106-110; held by Jews and Manichæans, 110-112; exegetically applied by Origen 112-114; taught by the Greeks, 114; in mythology and folklore, 117; basis of zoölatry and cannibalism, 118; as applied metaphor, 121; predestination explained by, 112, 113, 122, 129-132; individual extinction the final aim of, 127, 128; from an ethical point of view, 130; the embodiment of cherished propensities, 121, 131; the spiritual counterpart of metamorphosis, 132; as a code of morals in relation to the lower animals, 135-138; correspondence of the doctrine of evolution to, 162-164, 166.

Meyer, Hans, on driver ants, 247.
Milton, his Stygian metaphysicians, 130, 131; on man as God's master work, 193.
Missing link, probable language of the, 316.
Moebius, Dr., on Amtsberg's experiment with a pike, 186, 188.
Moggridge, on trapdoor spiders, 258.
Moleschott, 137.
Molothrus, its essays at nest-building, 199.
Monacella, St., protectress of hares, 154.
Monboddo, Lord, his theory of the origin of language, 290.
Monkeys, tool-using, 259, 261, 262; as miners, 263; use of fire, 264; reasoning from cause to effect by, 265, 266, 299; ability to count, 286; language of, 272, 314-317, 319, 324, 329-332; plastic period of, 337; musical concerts of, 269, 343, 349.
More, Henry, on four dimensions, 124.
Morioris, doctor revered as a god by, 352.
Mormons, sacredness of agriculture taught by, 72.
Moss, Capt. E., his monkeys as miners, 263.
Mother, descent from the, 26.
Müller, Max, on language as our Rubicon, 271-273; on roots as ultimate facts in language, 273; Sanskrit examples of the formation of roots, 274-276, 278, 281, 282, 303, 304; concomitant and significant clamour, 283, 305, 310; on the ability of animals to count and form general concepts, 281, 282, 285, 286; on interjections and exclamations, 283, 287; on a mere " fold of the brain," 295; on

the evolution of the eye, ear, and brain, 297; on "nursery philology," 301, 312; Noiré's theory indorsed by, 303, 307; on the Homo alalus, 304, 305; on the bow-wow, pooh-pooh, and yo-he-ho theories, 307; on philology in the menagerie, 312.
Munich, meaning of, 22; Thierschutzverein in, 149; former cruelty to animals in, 159.

Name day, importance of, 54.
Nanak, compared with Paul, 128.
Napoleon III, his appeal to tribal prejudice, 49.
Nation, modern conception of a, 46.
Negritos, as beasts of venery, 78.
Negroes, prejudice against, 77; analogies between monkeys and, 267, 268, 337.
Němici, Slavonic appellation of Germans, 24.
Neoherbartianism, 15.
Noolamarckian school of science 125.
Neoplatonism, 10, 137.
Neopythagoreanism, 10, 137.
Nicaise, his wonderful parrot, 326-328.
Nirvâna, final aim of Buddhism, 128, 129.
Noiré, on the origin of language, 303.
Nomadic tribes, transition to sedentary life, 42, 64, 70-73; ants, 247.
Norton, Allen H., logical faculty in his dog, 266.
Nursery, philology in the, 301, 312.
Nutrition, impulse of, 175-177.

Oken, his classification of animals, 178.
Opossum, power of simulation, 181.

Orang-outangs. See APES and MONKEYS.
Organisms, evolution of animated, 167, 171-174.
Orient, animal ethics in the, 9, 88, 136, 145; local significance of the term, 22; pantheism and atheism in the, 126; evolutional speculation in the, 162-166.
Origen, against Celsus, 10; on the purpose of animals, 10; metempsychosis the basis of his exegesis, 112, 113.
Orioles, skill in nest-building, 200, 201, 339.
Orpheus, reborn as a swan, 114.
Ovid, on the erect posture of man, 192; in the Pontus, 271.
Owl, mental horizon of the, 194.

Paley, his definition of virtue, 92.
Panis, his golden pheasants, 145.
Panjab, 68.
Panjara Pol, described, 139-142, 143.
Panpsychism, as the basis of animal rights, 137.
Panslavism and Panteutonism, 50.
Pantheism, 126, 129.
Pantomime, as a means of expression, 291-293.
Parrots, use of articulate sounds by, 271; remarkable, 325-327.
Parsi, classification of animals by the, 136.
Parson, dilemma of a German, 149.
Pasturage, fatal to hunting tribes, 71.
Paśu, meaning of, 7.
Patriotism, a survival of tribalism, 25; barbarous exhibitions of, 34-38, 40; Dr. Johnson's definition of, 228.
Paul, his influence on Christianity, 58; compared with Nanak, 128;

his doctrine of election, 130; on soul and body, 133; compared with Buddha, 164.
Paulsen, 187.
Penguin, erect posture of the, 194.
Perception, impulse of, 175-181.
Peter, on brute beasts, 152.
Pfeifer, his musically trained sparrow, 340.
Pharisees, as proselyters, 57.
Phenacodus Primævus, progenitor of hooked or clawed animals, 281.
Physical organs, their relation to mental development, 15-18.
Pig, mental deterioration through domestication, 217; as a hunter, 218.
Pike, Amtberg's experiment with a, 186-188.
Pithecoid man, bodily aids to the mental growth of, 17.
Pius IX, his dictum concerning animals, 160.
Plants, function of colour and odour in, 87.
Plato, on pure soul, 121; on the origin of *a priori* ideas, 123; on pre-existence, 125; on the song of the cicada, 342.
Plautus, Latin called barbarous by, 31.
Plimsoll, Samuel, on cattle ships, 101.
Pliny, on the incendiary bird, 265.
Plotinus, 10, 137.
Plutarch, 9; on religion as the foundation of the state, 55.
Poe, E. A., quoted, 85.
Points of the compass, 20, 22.
Polyandry, survivals of, 31, 41.
Polynesians, as affectionate parricides, 108.
Popes, Spanish bullfights and the, 161.

Porphyrius, 9, 137; on the language of animals, 293.
Pound, purpose of the, 144.
Prantl, Prof., on the mental operations of animals, 222-225, 265; denies "time-sense" to animals, 223; weak point of his speculations, 225; denies use of tools and fire to animals, 257, 264; on the art impulse in animals, 224, 334.
Prehension, as an aid to comprehension, 195.
Preyer, Prof., on animal simulation, 181.
Primitive man, aliens and animals in the eyes of, 4-8; limited world of, 20-22; idea of retributive justice entertained by, 37; tribal marriage of, 31, 50; his belief in the transmigration of souls, 107-109, 119.
Property, origin of the conception of personal, 176 · soko's idea of, 269.
Protista, nature of, 171, 172.
Protoplasm, evolution of organisms out of, 173.
Psophos, distinction between phôné and, 309.
Psychology, comparative, 2, 17, 162; scholastic, 2, 3, 220; Neoherbartian, 15; anthropocentric, 83. See ANIMALS and ZOÖPSYCHOLOGY.
Public opinion, wholesome influence of, 80.
Pythagoras, his memory of pre-existence, 114, 125, 137; fish released by, 154.

"Quack," an onomatopoetic root in child language, 276, 277, 279, 280.
Quadrumans, their advantage over quadrupeds, 336.

Quaquas, animal worship by, 119.
Quatrefages, on the religious sentiment in animals, 351.

Racine, on man's heavenward look, 193.
Radeau, R., on the language of animals, 323; Mersenne's theory of language rejected by, 324.
Rats, benevolence in, 258.
Rattlesnake, suicide of a, 227.
Ravens, infliction of punishment by, 234; use of fire by, 264; use of articulate speech by, 300.
Reason, boundaries of instinct and, 6, 95, 167, 170; in animals, see ANIMALS.
Reclam, Prof., on a musical spider, 341.
Relationship, real and ritual, 54.
Religion, moral influence of, 9; its value in early society, 55; in animals, see ANIMALS; as a revelation to man, 357, 358.
Richard, Jules, on simian speech, 324.
Rickaby, Rev. Joseph, animals not autocentric, 2.
Rivâyats, cited, 64.
Rochefort, termites in, 211.
Romanes, on intelligence in the amœba, 17; on tool-using animals, 259; on child language, 276; a chimpanzee's knowledge of numeration, 286; a speaking terrier, 300; on the Homo alalus, 305; example of religious awe in a Skye terrier, 355.
Rome, urbs et orbis, 21.
Roots of language. See MAX MÜLLER.
Rosegger, on anarchism in the hives, 240.
Rostan, Prof., his attack of aphasia, 298.

Rotundity of the earth, and the rights of man, 21.

Sacy, Silvestre de, on the study of Sanskrit, 318.
Sad-dar, cited, 63.
Sadducees, 57, 127.
Saints, their relations to animals, 152–156.
Sakarbai, Lady, her hospital for animals, 142.
Samodershez, title of Czar as tribal sovereign, 45.
Sanhedrin, 57.
Sanskrit, formation of roots in, 274, 278, 281; alleged superiority of, 284; as a fabrication of Brahmans, 318.
Satar, 64.
Savages, tribal spirit of, 2, 21; treatment of old and infirm, 139; ability to count and distinguish colours, 287, 351. See PRIMITIVE MAN.
Sayce, Prof., on Bushman speech, 301.
Schabbesgoi, 76.
Scheitling, on canine love of music, 344.
Schelling, on religion as the cement of society, 54.
Schiller, on love of nature by the Greeks, 190; on art as the prerogative of man, 334; on the play impulse in animals, 346.
Schlagintweit, on elephants as dam-builders, 259, 260.
Schleiermacher, 137.
Schneider, G. H., his four categories of mental impulses, 175; his psychological classification of animals, 178; on the impulses of food-storing animals, 179; on the stargazer, 185.
Schomburgk, Dr., simian reasoning from cause to effect, 265.

Schopenhauer, on a radical defect in Judaism and Christianity, 88; panpsychism, 137; on men as devils, 147.
Schutz, Leopold, irrationality of brutes a dogma of the Church, 96; animals as puppets, 219, 220.
Scorpions, suicide of, 227.
Scripture, place of animals in, 88: kindness to animals inculcated in passages from, 149, 150.
Scythians, their language like the chatter of cranes, 293.
Sea anemone, its relation to the hermit crab, 184.
Sea pudding, as gourmand, 170.
Second sight, ascribed to animals, 357.
Sedley, on St. Peter's crime, 161.
Semon, Richard, on Australian tribes, 287.
Senegambia, termite mounds in, 212.
Sensation, impulse of, 175, 177-181.
Sentinels, posted by animals, 182, 183.
Shadow-bird, as dancer, 350.
Shakespeare, quoted, 132, 317.
Shelley, on man's supremacy, 89.
Shylock, as typical Hebrew, 76.
Sikh, prophet, 128.
Silius Italicus, on man's upward look, 193.
Simmins, S., on "sweating" bees, 202-205.
Sioux, idea of retributive justice, 37.
Sister, marriage of own, 51, 52.
Slavonic, meaning of, 24.
Smeathman, on termite soldiers and workers, 210.
Smith, William B., his club-wielding donkey, 259.
Smith, Sydney, on brute souls as immortal, 94.

Socrates, on pre-existence, 114; Xantippe's grievance against, 176.
Soko, characteristics of the, 267-269, 343.
Solomon, his cynical view of life, 147, 148; on ants, 254.
Song. See BIRDS.
Sophocles, his use of phôné, 309.
Souls, congeniousness of, 15; influence of bodily conditions on, 16-18; immortality of brute, 93-96; transmigration of, 106 (see METEMPSYCHOSIS); origin of the belief in the immortality of, 177.
Spain, animals' rights not recognized in, 14, 161.
Sparrows, maternal training of, 189; improvements in nest-building made by, 202; musical education of, 340.
Spartans, guest-hating, 31.
Spencer, Baldwin, on the musical tones of the Australian spider, 342.
Spencer, Herbert, on the play impulse in animals, 346; on religious awe and propitiation in animals, 353, 354.
Spentô-mainyush, 8.
Spiders, their love of music, 340-342.
Spiegel, 62, 65.
Spinoza, on the indestructibility of mind, 123; as lens-grinder and philosopher, 176; on benevolence in animals, 228.
Spirit. See SOULS.
Squirrels, their impulses to action, 179; as explorers, 182.
Stargazer, as angler, 184.
Steinthal, on birds and brutes, 193, 194; on the origin of language, 305, 306.
Steward, Dugald, on the adaptabil-

ity of animals and its value to man, 98.
Stoics, liberal philosophy of the, 28, 137.
Storks, their conjugal relations and courts of justice, 230–234.
Suicide, by reptiles, 227.
Súnîsâr, an Indian atheistic poem, 127.
Śûnyavâdinaḥ, 127.
Svoboda, quoted, 177.
Swallows, their recent improvements in nest-building, 200.
Swedenborg, his vision, 120.
Swine, scientific breeding of, 214. See Pig.
Switzerland, clannish spirit in, 34, 36.
Symbol, as a token of hospitality, 32; Christian, 56.

Tailor-bird, its progress in artistic skill, 201.
Tait, Lawrence, on drunken wasps, 242.
Tamar, her relations to Amnon, 51.
Taylor, Father, on ruffianism, 147.
Teleology, absurdities of anthropocentric, 83–88.
Tennent, Sir James Emerson, on the Veddahs of Ceylon, 289, 290.
Terence, quoted, 24.
Termites, habitations of, 207–210, 212; the "royal residence" of, 208; engineering skill of, 209; destructiveness of, 211; proportion of soldiers to workers, 210, 211.
Tesseres hospitales, 32.
Theft, as a tribal virtue, 25.
Theocritus, on the soul of the Nemean lion, 93.
Theodicy, untenableness of orthodox, 104, 131.
Theophrastus, 9, 137.

Thibetans, polyandry of the, 31.
Thibo, King, his white elephant, 143.
Thomas Aquinas, his philosophy indorsed by the Catholic Church, 3; head of mediæval scholiasts, 12; his doctrine of objective necessity, 130; on the beast soul, 134.
Thoreau, his sympathy with animals, 159.
Thought, impulse of, 175–181.
Thracians, their contempt for husbandry, 43; their language like the chatter of cranes, 293.
Titmouse, skill of the, 339.
Tooke, Horne, on interjections, 283 285.
Tools, used by animals, 257–263.
Torres, Countess de la, her asylum for cats, 144.
Totemism, origin and survivals of, 6.
Treiber, Rev. A., his musical poodle, 344.
Trent, Council of, 54.
Tribe, ethics of the, 24, 73; genesis of the, 41–43; tribal superseded by territorial sovereignty, 42–46; marriage within the, 50–52; religion of the, 55–58.
Trousseau, on aphasia, 298.
Tumblebugs, insectile and human, 180.
Tycho de Brahe, his belief in astrology, 85.
Tyrell, Rev. George, his scholastic psychology, 3.

Umber, as a dancer, 350.
United States, national stronger than race feeling in the, 46; right of expatriation asserted by the, 47; legislation inconsistent with this right, 48.
Usury, primitive notions concern-

ing, 73; Cato on, 74; exacted by Jews from Gentiles, 74; laws regulating and punishing, 75.

Vaccination, denounced as impious, 84.
Vedânta, influence of the, 126.
Vedas, religion of the, 67.
Veddahs, not degraded but undeveloped, 290.
Veland, represented as lame, 44.
Vendidâd, dualism of the, 8, 59; sacredness of agriculture in the, 60-67.
Vinci, Leonardo da, his kindness to birds, 154.
Virâf, Ardâ, his polygamy and incest, 51.
Vital force, artificial products of, 174.
Vogt, Carl, on the conjugal infidelity of storks, 232; on the human and simian brain, 296.
Vogtleute, 36.
Vôhumanô, the Good Mind, 8, 62, 68.
Vulcan, lameness of, 44.

Wagtails, nests of, 202.
Waitz, Theodor, on barbarized Europeans, 79.
Wallace, A. R., on simian infancy, 335.
Wallace, D. Mackenzie, on the Cossacks, 71.
Walton, Izaak, on the nightingale's song, 340.
Wasps, effect of alcohol on, 242.
Wattenwyl, Brunner von, on the music of the cicada, 343.
Wayland, Dr., on duties to animals, 98.
Weaver bird, skill of the, 339.
Weir, James, battle of ants described by, 248; on the vocal organs of ants, 311.
Wenzel, G. I., his studies of animal speech, 320-323; conversation of foxes overheard by, 320; his alphabet of animal speech, 321; knowledge of human language acquired by dogs, 321, 322.
Wert, E. W., cited, 64.
Weygandt, on "milking" bees, 241.
Whitney, W. D., on language as a social institution, 288; on mispronunciation and false accent, 309; on the synergistic theory, 310.
Wildfangsrecht, 36.
Wilks, Dr., on knowledge of the fine arts as distinctively human, 333.
Williams, Monier, on hospitals for animals in India, 139-141.
Wives, three kinds of Persian, 64.
Wöhler, urea chemically produced by, 174.
Woman, transmission of race qualities through, 27; the first agriculturist, 43.
Wooing, primitive method of, 50.
Wordsworth, quoted, 317.
Wundt, Wilhelm, 137; on differences of degree between man and brute, 170, 171; on apian tradition, 244, note; on animal language, 293.

Xantippe, the domestic character of, 176; New England parallels to, 177.
Xenarchos, on the enviable muteness of the female cicada, 342.

Yâjñavalkya, definition of fate by, 122.
Yima, enlargement of the earth by, 64-66.
Yukan, Persian child-wife, 64.

Zarathustra, his classification of animals, 8; social reformation by,

58; his teachings, 59–63, 66–68; holiness of husbandry proclaimed by, 60, 72.

Zebras, ostriches used as sentries by, 183.

Zoölatry, basis of, 6; ethical influence of, 7; its relation to metempsychosis, 118; survivals of, 6, 119

Zoöphily, ethics of, 3; scientific basis of, 96.

Zoöpsychology, a branch of comparative psychology, 17; ethical influence of, 96; metaphysical barriers between man and beast gradually removed by, 162–164, 171. See ANIMALS.

THE END.

D. APPLETON AND COMPANY'S PUBLICATIONS.

THE ANTHROPOLOGICAL SERIES.

NOW READY.

THE BEGINNINGS OF ART. By ERNST GROSSE, Professor of Philosophy in the University of Freiburg. A new volume in the Anthropological Series, edited by Professor Frederick Starr. Illustrated. 12mo. Cloth, $1.75.

"This book can not fail to interest students of every branch of art, while the general reader who will dare to take hold of it will have his mind broadened and enriched beyond what he would conceive a work of many times its dimensions might effect."—*Brooklyn Eagle.*

"The volume is clearly written, and should prove a popular exposition of a deeply interesting theme."—*Philadelphia Public Ledger.*

WOMAN'S SHARE IN PRIMITIVE CULTURE. By OTIS TUFTON MASON, A. M., Curator of the Department of Ethnology in the United States National Museum. With numerous Illustrations. 12mo. Cloth, $1.75.

"A most interesting *résumé* of the revelations which science has made concerning the habits of human beings in primitive times, and especially as to the place, the duties, and the customs of women."—*Philadelphia Inquirer.*

THE PYGMIES. By A. DE QUATREFAGES, late Professor of Anthropology at the Museum of Natural History, Paris. With numerous Illustrations. 12mo. Cloth, $1.75.

"Probably no one was better equipped to illustrate the general subject than Quatrefages. While constantly occupied upon the anatomical and osseous phases of his subject, he was none the less well acquainted with what literature and history had to say concerning the pygmies. . . . This book ought to be in every divinity school in which man as well as God is studied, and from which missionaries go out to convert the human being of reality and not the man of rhetoric and text-books."—*Boston Literary World.*

THE BEGINNINGS OF WRITING. By W. J. HOFFMAN, M. D. With numerous Illustrations. 12mo. Cloth, $1.75.

This interesting book gives a most attractive account of the rude methods employed by primitive man for recording his deeds. The earliest writing consists of pictographs which were traced on stone, wood, bone, skins, and various paperlike substances. Dr. Hoffman shows how the several classes of symbols used in these records are to be interpreted, and traces the growth of conventional signs up to syllabaries and alphabets—the two classes of signs employed by modern peoples.

IN PREPARATION.

THE SOUTH SEA ISLANDERS. By Dr. SCHMELTZ.

THE ZUÑI. By FRANK HAMILTON CUSHING.

THE AZTECS. By Mrs. ZELIA NUTTALL.

D. APPLETON AND COMPANY, NEW YORK.

D. APPLETON AND COMPANY'S PUBLICATIONS.

THE CRIMINOLOGY SERIES.
Edited by W. Douglas Morrison.

CRIMINAL SOCIOLOGY. By Professor E. Ferri. 12mo. Cloth, $1.50.

"A most valuable book. It is suggestive of reforms and remedies, it is reasonable and temperate, and it contains a world of information and well-arranged facts for those interested in or merely observant of one of the great questions of the day."—*Philadelphia Public Ledger.*

"The scientist, the humanitarian, and the student will find much to indorse and to adopt, while the layman will wonder why such a book was not written years ago."—*Newark Advertiser.*

THE FEMALE OFFENDER. By Professor Lombroso. Illustrated. 12mo. Cloth, $1.50.

"'The Female Offender' must be considered as a very valuable addition to scientific literature. . . . It is not alone to the scientist that the work will recommend itself. The humanitarian, anxious for the reform of the habitual criminal, will find in its pages many valuable suggestions."—*Philadelphia Item.*

"This work will undoubtedly be a valuable addition to the works on criminology, and may also prove of inestimable help in the prevention of crime."—*Detroit Free Press.*

"The book is a very valuable one, and admirably adapted for general reading."—*Boston Home Journal.*

"This book will probably constitute one of the most valuable contributions in aid of a nineteenth-century sign of advanced civilization known as prison reform."—*Cincinnati Times-Star.*

"There is no book of recent issue that bears such important relation to the great subject of criminology as this book."—*New Haven Leader.*

OUR JUVENILE OFFENDERS. By W. Douglas Morrison, author of "Jews under the Romans," etc. 12mo. Cloth, $1.50.

"An admirable work on one of the most vital questions of the day. . . . By scientists, as well as by all others who are interested in the welfare of humanity, it will be welcomed as a most valuable and a most timely contribution to the all-important science of criminology."—*New York Herald.*

"Of real value to scientific literature. In its pages humanitarians will find much to arrest their attention and direct their energies in the interest of those of the young who have gone astray."—*Boston Daily Globe.*

IN PREPARATION.

CRIME A SOCIAL STUDY. By Professor Joly.

CRIMINAL ANTHROPOLOGY. By Dr. J. Dallemagne.

D. APPLETON AND COMPANY, NEW YORK.

D. APPLETON AND COMPANY'S PUBLICATIONS.

DEGENERATION. By Professor MAX NORDAU. Translated from the second edition of the German work. 8vo. Cloth, $3.50.

"A powerful, trenchant, savage attack on all the leading literary and artistic idols of the time by a man of great intellectual power, immense range of knowledge, and the possessor of a lucid style rare among German writers, and becoming rarer everywhere, owing to the very influences which Nordau attacks with such unsparing energy, such eager hatred."—*London Chronicle.*

"The wit and learning, the literary skill and the scientific method, the righteous indignation, and the ungoverned prejudice displayed in Herr Max Nordau's treatise on 'Degeneration,' attracted to it, on its first appearance in Germany, an attention that was partly admiring and partly astonished."—*London Standard.*

"Let us say at once that the English-reading public should be grateful for an English rendering of Max Nordau's polemic. It will provide society with a subject that may last as long as the present Government. . . . We read the pages without finding one dull, sometimes in reluctant agreement, sometimes with amused content, sometimes with angry indignation."—*London Saturday Review.*

"Herr Nordau's book fills a void, not merely in the systems of Lombroso, as he says, but in all existing systems of English and American criticism with which we are acquainted. It is not literary criticism, pure and simple, though it is not lacking in literary qualities of a high order, but it is something which has long been needed. . . . A great book, which every thoughtful lover of art and literature and every serious student of sociology and morality should read carefully and ponder slowly and wisely."—*Richard Henry Stoddard, in the Mail and Express.*

"The book is one of more than ordinary interest. Nothing just like it has ever been written. Agree or disagree with its conclusions, wholly or in part, no one can fail to recognize the force of its argument and the timeliness of its injunctions."—*Chicago Evening Post.*

GENIUS AND DEGENERATION. A Study in Psychology. By Dr. WILLIAM HIRSCH. Translated from the second edition of the German work. Uniform with "Degeneration." Large 8vo. Cloth, $3.50.

"The first intelligent, rational, and scientific study of a great subject. . . . In the development of his argument Dr. Hirsch frequently finds it necessary to attack the positions assumed by Nordau and Lombroso, his two leading adversaries. . . . Only calm and sober reason endure. Dr. Hirsch possesses that calmness and sobriety. His work will find a permanent place among the authorities of science."—*N. Y. Herald.*

"Dr. Hirsch's researches are intended to bring the reader to the conviction that 'no psychological meaning can be attached to the word genius.' . . . While all men of genius have common traits, they are not traits characteristic of genius; they are such as are possessed by other men, and more or less by all men. . . . Dr. Hirsch believes that most of the great men, both of art and of science, were misunderstood by their contemporaries, and were only appreciated after they were dead."—*Miss J. L. Gilder, in the Sunday World.*

"'Genius and Degeneration' ought to be read by every man and woman who professes to keep in touch with modern thought. It is deeply interesting, and so full of information that by intellectual readers it will be seized upon with avidity."—*Buffalo Commercial.*

D. APPLETON AND COMPANY, NEW YORK.

D. APPLETON AND COMPANY'S PUBLICATIONS.

PIONEERS OF EVOLUTION, from Thales to Huxley. By EDWARD CLODD, President of the Folk-Lore Society; Author of "The Story of Creation," "The Story of 'Primitive' Man," etc. With Portraits. 12mo. Cloth, $1.50.

"The mass of interesting material which Mr. Clodd has got together and woven into a symmetrical story of the progress from ignorance and theory to knowledge and the intelligent recording of fact is prodigious. . . . The 'goal' to which Mr. Clodd leads us in so masterly a fashion is but the starting point of fresh achievements, and, in due course, fresh theories. His book furnishes an important contribution to a liberal education."—*London Daily Chronicle.*

"We are always glad to meet Mr. Clodd. He is never dull; he is always well informed, and he says what he has to say with clearness and precision. . . . The interest intensifies as Mr. Clodd attempts to show the part really played in the growth of the doctrine of evolution by men like Wallace, Darwin, Huxley, and Spencer. . . . We commend the book to those who want to know what evolution really means."—*London Times.*

"This is a book which was needed. . . . Altogether, the book could hardly be better done. It is luminous, lucid, orderly, and temperate. Above all, it is entirely free from personal partisanship. Each chief actor is sympathetically treated, and friendship is seldom or never allowed to overweight sound judgment."—*London Academy.*

"We can assure the reader that he will find in this work a very useful guide to the lives and labors of leading evolutionists of the past and present. Especially serviceable is the account of Mr. Herbert Spencer and his share in rediscovering evolution, and illustrating its relations to the whole field of human knowledge. His forcible style and wealth of metaphor make all that Mr. Clodd writes arrestive and interesting."—*London Literary World.*

"Can not but prove welcome to fair-minded men. . . . To read it is to have an object-lesson in the meaning of evolution. . . . There is no better book on the subject for the general reader. . . . No one could go through the book without being both refreshed and newly instructed by its masterly survey of the growth of the most powerful idea of modern times."—*The Scotsman.*

D. APPLETON AND COMPANY, NEW YORK.

D. APPLETON & CO.'S PUBLICATIONS.

THE SYNTHETIC PHILOSOPHY OF HERBERT SPENCER. In ten volumes. 12mo. Cloth, $2.00 per volume. The titles of the several volumes are as follows:

(1.) FIRST PRINCIPLES.
    I. The Unknowable.   II. Laws of the Knowable.

(2.) THE PRINCIPLES OF BIOLOGY. Vol. I.
    I. The Data of Biology.   II. The Inductions of Biology.
    III. The Evolution of Life.

(3.) THE PRINCIPLES OF BIOLOGY. Vol. II.
    IV. Morphological Development.   V. Physiological Development.
    VI. Laws of Multiplication.

(4.) THE PRINCIPLES OF PSYCHOLOGY. Vol. I.
    I. The Data of Psychology.   III. General Synthesis.
    II. The Inductions of Psychology.   IV. Special Synthesis.
    V. Physical Synthesis.

(5.) THE PRINCIPLES OF PSYCHOLOGY. Vol. II.
    VI. Special Analysis.   VIII. Congruities.
    VII. General Analysis.   IX. Corollaries.

(6.) THE PRINCIPLES OF SOCIOLOGY. Vol. I.
    I. The Data of Sociology.   II. The Inductions of Sociology.
    III. The Domestic Relations.

(7.) THE PRINCIPLES OF SOCIOLOGY. Vol. II.
    IV. Ceremonial Institutions.   V. Political Institutions.

(8.) THE PRINCIPLES OF SOCIOLOGY. Vol. III.
    VI. Ecclesiastical Institutions.   VII. Professional Institutions.
    VIII. Industrial Institutions.

(9.) THE PRINCIPLES OF ETHICS. Vol. I.
    I. The Data of Ethics.   II. The Inductions of Ethics.
    III. The Ethics of Individual Life.

(10.) THE PRINCIPLES OF ETHICS. Vol. II.
    IV. The Ethics of Social Life: Justice.
    V. The Ethics of Social Life: Negative Beneficence.
    VI. The Ethics of Social Life: Positive Beneficence.

DESCRIPTIVE SOCIOLOGY. *A Cyclopædia of Social Facts.* Representing the Constitution of Every Type and Grade of Human Society, Past and Present, Stationary and Progressive. By HERBERT SPENCER. Eight Nos., Royal Folio.

No. I. ENGLISH	$4 00
No. II. MEXICANS, CENTRAL AMERICANS, CHIBCHAS, and PERUVIANS	4 00
No. III. LOWEST RACES, NEGRITO RACES, and MALAYO-POLYNESIAN RACES	4 00
No. IV. AFRICAN RACES	4 00
No. V. ASIATIC RACES	4 00
No. VI. AMERICAN RACES	4 00
No. VII. HEBREWS and PHŒNICIANS	4 00
No. VIII. FRENCH (Double Number)	7 00

D. APPLETON AND COMPANY, NEW YORK.

D. APPLETON & CO.'S PUBLICATIONS.

NEW EDITION OF PROF. HUXLEY'S ESSAYS.

COLLECTED ESSAYS. By THOMAS H. HUXLEY. New complete edition, with revisions, the Essays being grouped according to general subject. In nine volumes, a new Introduction accompanying each volume. 12mo. Cloth, $1.25 per volume.

- VOL. I.—METHOD AND RESULTS.
- VOL. II.—DARWINIANA.
- VOL. III.—SCIENCE AND EDUCATION.
- VOL. IV.—SCIENCE AND HEBREW TRADITION.
- VOL. V.—SCIENCE AND CHRISTIAN TRADITION.
- VOL. VI.—HUME.
- VOL. VII.—MAN'S PLACE IN NATURE.
- VOL. VIII.—DISCOURSES, BIOLOGICAL AND GEOLOGICAL.
- VOL. IX.—EVOLUTION AND ETHICS, AND OTHER ESSAYS.

"Mr. Huxley has covered a vast variety of topics during the last quarter of a century. It gives one an agreeable surprise to look over the tables of contents and note the immense territory which he has explored. To read these books carefully and studiously is to become thoroughly acquainted with the most advanced thought on a large number of topics."—*New York Herald.*

"The series will be a welcome one. There are few writings on the more abstruse problems of science better adapted to reading by the general public, and in this form the books will be well in the reach of the investigator. . . . The revisions are the last expected to be made by the author, and his introductions are none of earlier date than a few months ago [1893], so they may be considered his final and most authoritative utterances."—*Chicago Times.*

"It was inevitable that his essays should be called for in a completed form, and they will be a source of delight and profit to all who read them. He has always commanded a hearing, and as a master of the literary style in writing scientific essays he is worthy of a place among the great English essayists of the day. This edition of his essays will be widely read, and gives his scientific work a permanent form."—*Boston Herald.*

"A man whose brilliancy is so constant as that of Prof. Huxley will always command readers; and the utterances which are here collected are not the least in weight and luminous beauty of those with which the author has long delighted the reading world."—*Philadelphia Press.*

"The connected arrangement of the essays which their reissue permits brings into fuller relief Mr. Huxley's masterly powers of exposition. Sweeping the subject-matter clear of all logomachies, he lets the light of common day fall upon it. He shows that the place of hypothesis in science, as the starting point of verification of the phenomena to be explained, is but an extension of the assumptions which underlie actions in every-day affairs; and that the method of scientific investigation is only the method which rules the ordinary business of life."—*London Chronicle.*

New York: D. APPLETON & CO., 72 Fifth Avenue.

D. APPLETON & CO.'S PUBLICATIONS.

PIONEERS OF SCIENCE IN AMERICA.

Sketches of their Lives and Scientific Work. Edited and revised by WILLIAM JAY YOUMANS, M.D. With Portraits. 8vo. Cloth, $4.00.

Impelled solely by an enthusiastic love of Nature, and neither asking nor receiving outside aid, these early workers opened the way and initiated the movement through which American science has reached its present commanding position. This book gives some account of these men, their early struggles, their scientific labors, and, whenever possible, something of their personal characteristics. This information, often very difficult to obtain, has been collected from a great variety of sources, with the utmost care to secure accuracy. It is presented in a series of sketches, some fifty in all, each with a single exception accompanied with a well-authenticated portrait.

"Fills a place that needed filling, and is likely to be widely read."—*N. Y. Sun.*

"It is certainly a useful and convenient volume, and readable too, if we judge correctly of the degree of accuracy of the whole by critical examination of those cases in which our own knowledge enables us to form an opinion. . . . In general, it seems to us that the handy volume is specially to be commended for setting in just historical perspective many of the earlier scientists who are neither very generally nor very well known."—*New York Evening Post.*

"A wonderfully interesting volume. Many a young man will find it fascinating. The compilation of the book is a work well done, well worth the doing."—*Philadelphia Press.*

"One of the most valuable books which we have received."—*Boston Advertiser.*

"A book of no little educational value. . . . An extremely valuable work of reference."—*Boston Beacon.*

"A valuable handbook for those whose work runs on these same lines, and is likely to prove of lasting interest to those for whom '*les documents humain*' are second only to history in importance—nay, are a vital part of history."—*Boston Transcript.*

"A biographical history of science in America, noteworthy for its completeness and scope. . . . All of the sketches are excellently prepared and unusually interesting."—*Chicago Record.*

"One of the most valuable contributions to American literature recently made. . . . The pleasing style in which these sketches are written, the plans taken to secure accuracy, and the information conveyed, combine to give them great value and interest. No better or more inspiring reading could be placed in the hands of an intelligent and aspiring young man."—*New York Christian Work.*

"A book whose interest and value are not for to-day or to-morrow, but for indefinite time."—*Rochester Herald.*

"It is difficult to imagine a reader of ordinary intelligence who would not be entertained by the book. . . . Conciseness, exactness, urbanity of tone, and interestingness are the four qualities which chiefly impress the reader of these sketches."—*Buffalo Express.*

"Full of interesting and valuable matter."—*The Churchman.*

New York: D. APPLETON & CO., 72 Fifth Avenue.

D. APPLETON & CO.'S PUBLICATIONS.

INTRODUCTION TO THE STUDY OF PHILOSOPHY. By WILLIAM T. HARRIS, LL. D., United States Commissioner of Education. Compiled and arranged by MARIETTA KIES. 12mo. Cloth, $1.50.

"Philosophy as presented by Dr. Harris gives to the student an interpretation and explanation of the phases of existence which render even the ordinary affairs of life in accordance with reason; and for the higher or spiritual phases of life his interpretations have the power of a great illumination. Many of the students are apparently awakened to an interest in philosophy, not only as a subject to be taken as a prescribed study, but also as a subject of fruitful interest for future years and as a key which unlocks many of the mysteries of other subjects pursued in a college course."—*From the Compiler's Preface.*

A HISTORY OF PHILOSOPHY IN EPITOME. By ALBERT SCHWEGLER. Translated from the first edition of the original German by JULIUS H. SEELYE. Revised from the ninth German edition, containing Important Additions and Modifications, with an Appendix, continuing the History in its more Prominent Lines of Development since the Time of Hegel, by BENJAMIN T. SMITH. 12mo. Cloth, $2.00.

"Schwegler's History of Philosophy is found in the hands of almost every student in the philosophical department of a German University, and is highly esteemed for its clearness, conciseness, and comprehensiveness. The present translation was undertaken with the conviction that the work would not lose its interest or its value in an English dress, and with the hope that it might be of wider service in such a form to students of philosophy here. It was thought especially that a proper translation of this manual would supply a want for a suitable text-book on this branch of study, long felt by both teachers and students in our American colleges."—*From the Preface.*

BIOGRAPHICAL HISTORY OF PHILOSOPHY, from its Origin in Greece down to the Present Day. By GEORGE HENRY LEWES. Two volumes in one. 8vo. Cloth, $3.50. Also in 2 vols., small 8vo. Cloth, $4.00.

"Philosophy was the great initiator of science. It rescued the nobler part of man from the dominion of brutish apathy and helpless ignorance, nourished his mind with mighty impulses, exercised it in magnificent efforts, gave him the unslaked, unslakable thirst for knowledge which has dignified his life, and enabled him to multiply tenfold his existence and his happiness. Having done this, its part is played. Our interest in it now is purely historical. The purport of this history is to show how and why the interest in philosophy has become purely historical."—*From the Introduction.*

New York: D. APPLETON & CO., 72 Fifth Avenue.

www.ingramcontent.com/pod-product-compliance
Lightning Source LLC
Chambersburg PA
CBHW051247300426
44114CB00011B/926